Laboratory Exercises in Mechatronics

Musa Jouaneh

*Department of Mechanical,
Industrial, & Systems Engineering
University of Rhode Island*

CENGAGE
Learning™

Australia • Brazil • Japan • Korea • Mexico • Singapore • Spain • United Kingdom • United States

CENGAGE
Learning™

Laboratory Exercises in Mechatronics
Musa Jouaneh

Publisher, Global Engineering:
Christopher M. Shortt

Acquisitions Editor: Swati Meherishi

Senior Developmental Editor:
Hilda Gowans

Editorial Assistant: Tanya Altieri

Team Assistant: Carly Rizzo

Marketing Manager: Lauren Betsos

Media Editor: Chris Valentine

Director, Content and Media
Production: Patricia M. Boies

Content Project Manager: Jennifer A.
Ziegler

Production Service: RPK Editorial
Services, Inc.

Copyeditor: Erin Wagner

Proofreader: Harlan James

Compositor: MPS Limited, a Macmillan
Company

Senior Art Director: Michelle Kunkler

Internal Designer: Juli Cook/Plan-
IT_Publishing/Carmela Periera

Cover Designer: Andrew Adams/
4065042 Canada Inc.

Cover Image:
© Raimundas/Shutterstock

Rights Acquisitions Specialist:
Sam Marshall

Text and Image Permissions Researcher:
Kristiina Paul

Senior First Print Buyer: Doug Wilke

For product information and technology assistance, contact us at
Cengage Learning Customer & Sales Support, 1-800-354-9706.

For permission to use material from this text or product,
submit all requests online at **www.cengage.com/permissions**.
Further permissions questions can be emailed to
permissionrequest@cengage.com.

Library of Congress Control Number: 2011934122

ISBN-13: 978-1-111-57025-5
ISBN-10: 1-111-57025-6

Cengage Learning
20 Channel Center Street
Boston, MA 02210
USA

Cengage Learning is a leading provider of customized learning solutions with office locations around the globe, including Singapore, the United Kingdom, Australia, Mexico, Brazil, and Japan. Locate your local office at **www.cengage.com/global**.

Cengage Learning products are represented in Canada by Nelson Education Ltd.

For your course and learning solutions, visit
www.cengage.com/engineering.

Purchase any of our products at your local college store or at our preferred online store **www.cengagebrain.com**.

Certain materials contained herein are reprinted with the permission of Microchip Technology Incorporated. No further reprints or reproductions may be made of said materials without Microchip Technology Inc.'s prior written consent.

Disclaimer

Publisher does not warrant or guarantee any of the products or experiments described herein or perform any independent analysis in connection with any of the product information contained herein. Publisher does not assume, and expressly disclaims, any obligation to obtain and include information other than that provided to it by the author. The reader is expressly warned to consider and adopt all safety precautions that might be indicated by the activities described herein and to avoid all potential hazards. By following the instructions contained herein, the reader willingly assumes all risks in connection with such instructions. The publisher makes no representations or warranties of any kind, including but not limited to, the warranties of fitness for particular purpose or merchantability, nor are any such representations implied with respect to the material set forth herein, and the publisher takes no responsibility with respect to such material. The publisher shall not be liable for any special, consequential, or exemplary damages resulting, in whole or part, from the readers' use of, or reliance upon, this material.

Printed at CLDPC, USA, 12-18

To the LORD who has done wonderful things in my life
and to my children Terra Marie, Gina Christie, and Charles Joseph

CONTENTS

1. No food or drinks are allowed in the lab.

2. Be alert and pay careful attention to what you are doing; don't get distracted.

3. Never handle electrical equipment with wet hands.

4. Check all connections (visually and using the multimeter) before you turn on the power supply, and do not apply a voltage higher than the rated voltage to the particular device you are using.

5. Never connect the positive power supply lead directly to ground. This will cause a short circuit.

6. Turn off immediately the power supply if you smell any burning. Do not turn the power back on until you have checked your circuit.

Instruments
- Multimeter (suggested GW GDM-8145)
- Variable power supply (suggested EXTECH 382213)
- Function generator (suggested Instek GFG-8215A)
- Oscilloscope (suggested TENMA 72-6810)

PC and Software
- Personal computer
- Data acquisition board (suggested Measurement Computing PCIM-DAS1602/16 with a 16-channel, 16-bit A/D; 2 D/A; 32 DIO; and 3 16-bit counters)
- Visual Basic Express 2010

Micro-controllers
- PICkit 2 or PICkit 3 programmer
- MPLAB (if using PICkit 3 programmer)
- Microchip Low Pin-Count Development board with PIC16F690 MCU (or Microchip PICDEM PIC18 Exploration board with PIC18F8722 MCU. This board has a built-in RS-232 connector which simplifies serial interfacing to a PC. It also allows easy access to many of the pins on the MCU)
- CSS, Inc. compiler for PIC MCU (PIC-C)

Miscellaneous
- BNC and banana type cables
- Breadboard
- Wires

This book contains mechatronics laboratory exercises that are designed to give the student a hands-on experience with applications of the concepts covered in the accompanying textbook, *Fundamentals of Mechatronics* by the same author. This book contains fourteen laboratory exercises plus a section that has a list of suggested extended or final projects. The order of the lab exercises corresponds with the main textbook sections relevant to these exercises. The first six laboratory exercises are designed to illustrate basic measurements, electrical circuits and electronic concepts. Exercises 7 through 14 focus on microcontrollers, timing and state-transition diagrams, sensors, stepper motors, and feedback control.

The level of difficulty of the exercises vary. Lab Exercises 7 through 14 tend to be more involved and will take longer time to do than the first six exercises (especially Lab 11, which deals with timing functions on a PC and state-transition diagrams). All of the given exercises will not be able to be covered in one semester (14 weeks), but are given such that the instructor has the choice of selecting exercises to give focus on particular concepts. Each laboratory has a set of structured, guided exercises plus one or more of less structure where the student has to come up with the details to perform the exercise. These less-structured exercises are indicated by an asterisk. The instructor can assign the less-structured exercises or skip them. The suggested projects section has a list of five detailed project ideas. These project ideas make use of the previously covered laboratory exercises.

Note that the author has tried to select low-cost and widely available components for almost all of the exercises. This allows instructors to provide a mechatronics laboratory experience for many students with a modest investment in components and minimal technician support. The author will provide at the text website an up-to-date part ordering list for the components needed for these exercises.

An Instructor's Solutions Manual for *Laboratory Exercises in Mechatronics* is available from the publisher on request. To request access to the solutions manual and additional course materials, please visit www.cengagebrain.com. At the cengagebrain.com home page, search for the ISBN of your title (from the back cover of your book) using the search box at the top of the page. This will take you to the product page where these resources can be found.

The author would like to acknowledge the many students who were enrolled in the mechatronics class at URI and who provided useful suggestions and comments which shaped the current manuscript. The author also would like to acknowledge James Byrnes, the electronic technician in the Mechanical, Industrial, and Systems Engineering Department at URI who helped in building and wiring some of the circuits used in this book. Also the author is grateful for the following reviewers for their comments and suggestions:

Alan A. Barhorst, Texas Tech University

Jordan M. Berg, Texas Tech University

William W. Clark, University of Pittsburgh

Burford Furman, San Jose State University

Hector Gutierrez, Florida Institute of Technology

Steve Hung, Clemson University

Marcia K. O'Malley. Rice University

and for the editorial staff at Cengage Learning and RPK Editorial Services.

Musa Jouaneh is a Professor of Mechanical Engineering in the Department of Mechanical, Industrial, and Systems Engineering at the University of Rhode Island (URI) where he has been working since 1990. He received his Bachelor of Science degree in Mechanical Engineering in 1984 from the University of Louisiana at Lafayette, and his Master and Doctorate degrees in Mechanical Engineering from the University of California at Berkeley in 1986 and 1989, respectively. His research interests include mechatronics and robotics with a particular interest in motion-control systems. He is the author or co-author of over 65 publications and holds two U.S. patents. He has served as a consultant to many companies in the Northeast and has received two URI College of Engineering Faculty Excellence Awards and the URI Foundation Teaching Excellence Award. Dr. Jouaneh is a member of American Society of Mechanical Engineers (ASME) and a senior member of Institute of Electrical and Electronic Engineers (IEEE).

Basic Measurements

After you complete this lab, you should be able to:

- Know how to operate basic electrical measuring instruments
- Describe the construction of a breadboard
- Know how to operate a function generator
- Explain how to operate an oscilloscope and how it can be used to perform time-based measurements
- Know how to read resistor values
- Know how to measure voltage and current

For this lab you need:

- Resistors: 1 kΩ, 5.1 kΩ, and 10 kΩ
- 0.1 μF capacitor
- 5 V light bulb
- Two SPDT switches

LAB BACKGROUND/INFORMATION:

In this lab, you will learn how to **construct simple electrical circuits**. The electrical circuits will be constructed on a breadboard, a description of which is given next. You also will use a **multimeter**, which is a device that combines several electrical meters, including: a voltmeter to measure voltage, an ammeter to measure current, and an ohmmeter to measure resistance. You also will be introduced to how to read resistor and capacitor values. You also will learn how to perform time-based measurements using a function generator and an oscilloscope.

BREADBOARD

A **breadboard** is a board that is used for the solderless connection of circuit components. Components (such as integrated circuit (IC) chips, resistors, and wires) are simply inserted into holes in the board. A typical breadboard is shown in Figure 1.1. Notice that the holes in the breadboard are connected differently, depending on the area of the board. In the *Bus strip* area, the holes in each strip are connected along the whole length of the board, while in the *Terminal strips* area, each of the five holes on each side of the notch are connected together.

MULTIMETER

A typical **multimeter** is shown in Figure 1.2. The leads to the multimeter are connected differently depending on what measurement needs to be performed. To

Figure 1.1

Breadboard
(Jouaneh, University of
Rhode Island.)

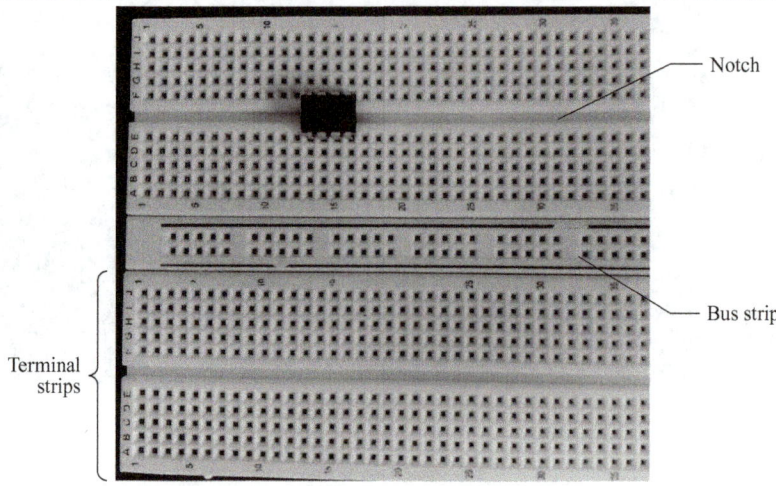

Notch

Bus strip

Terminal
strips

Figure 1.2

A multimeter
(Jouaneh, University of
Rhode Island.)

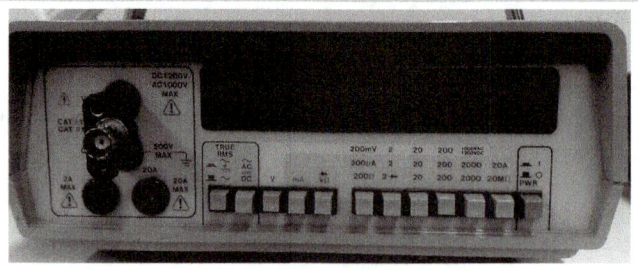

measure voltage and resistance, the measuring cable is connected to the upper (red) and to the middle (black) banana plugs (as shown in the figure). To **measure current**, the measuring cable is connected to one of the two lower (red) and to the middle (black) banana plugs.

FUNCTION GENERATOR

A **function generator** is a device that can generate various types of signals (such as triangular, rectangular, or sinusoidal). A typical function generator is shown in Figure 1.3.

Figure 1.3

Function generator
(Jouaneh, University of
Rhode Island.)

Offset knob

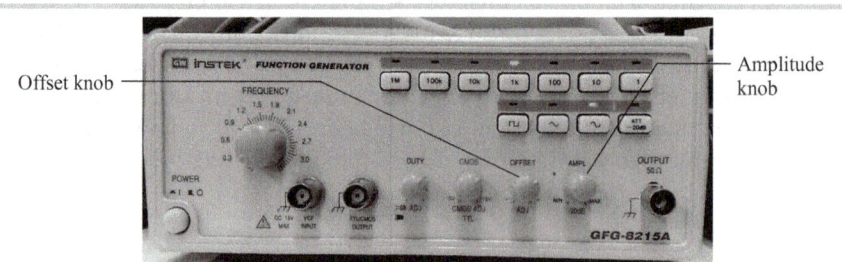

Amplitude
knob

In generating a signal from a function generator, four things have to be specified. These include the signal type, the signal frequency, the signal amplitude, and the signal offset. Each one is described here.

Signal Type: This refers to the signal shape (such as square or triangular). It is selected by pressing on one of the raised buttons that has the desired signal shape printed on it.

Signal Frequency: The signal frequency is selected by first selecting the signal frequency multiplier (i.e., 1, 10, 100, 1000, . . .) and then by adjusting the frequency knob. For example, to select a frequency of 2000 Hz, the frequency knob should be turned to the "2" mark and the 1000 multiplier button should be pressed down.

Signal Amplitude: This refers to the peak-to-peak signal value. It is adjusted by rotating the amplitude knob (AMPL).

Signal Offset: The signal offset adjusts the mean of the signal to a value other than zero by using the offset knob. For example, by setting the amplitude to 1 volt and using a signal offset of 0.5 volts, the peak voltage is 1 volt and the minimum voltage is zero.

OSCILLOSCOPE

An **oscilloscope** is a versatile laboratory device that is used to perform **time-based measurements** of signals. Most of the newer oscilloscopes are digital, which allows the signals to be saved. While oscilloscopes vary in the features they have, we will discuss some of the basic elements that are common to many of them. These include the display area, the voltage-per-division knob (VOLTS/DIV), and the time-per-division knob (TIME/DIV). Refer to Figure 1.4, which shows these elements on an oscilloscope made by TENMA®.

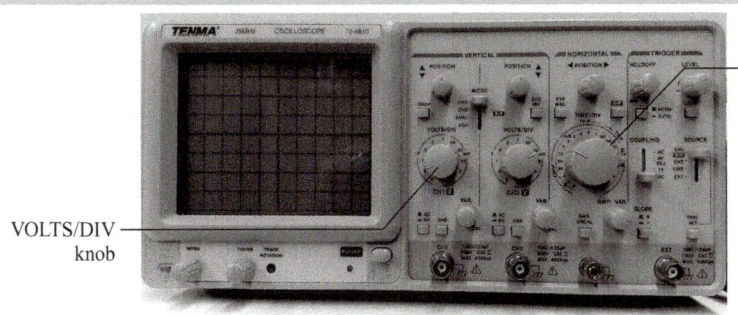

VOLTS/DIV knob

TIME/DIV knob

Figure 1.4

Front panel of a typical oscilloscope
(Jouaneh, University of Rhode Island.)

The **display area** is a rectangular window that displays the signals connected to the oscilloscope. One or two signals can be connected to the oscilloscope. One of the signal inputs is labeled CH1, while the other is labeled CH2. Depending on the input **mode** setting, one or both signal inputs can be displayed simultaneously. The different possibilities are

CH1: Only CH1 input is displayed.

CH2: Only CH2 input is displayed.

ADD: The CH1 and CH2 signals are added and displayed as a single signal.

DUAL: Both CH1 and CH2 inputs are displayed as separate signals.

Note that normally one or both input signals are displayed with respect to time, unless the **x-y** selection has been made. In that case, the two signals are displayed with respect to each other.

The **VOLTS/DIV knob** controls the scaling of the display area. For example, at the 1 volts/division setting, the height of each row in the display area is 1 V. Since typically the display area is divided into 8 rows, a signal with a peak-to-peak voltage greater than 8 V will be truncated. The location of the signal in the display area is adjusted by rotating the vertical and horizontal position knobs.

In a similar fashion, the **TIME/DIV knob** controls the horizontal scaling of the display area. The higher the frequency of the displayed signal, the smaller the value that the time/division knob should be set to.

The channel (CH1 and CH2) inputs have a BNC type connector (see Appendix A) to facilitate the attachment of signals.

Figure 1.5

Resistor color bands

a b c Tol

RESISTOR COLOR CODES

Resistors have color codes that are used to determine the resistor value. Typically, **four color bands** are shown on the resistor as shown in Figure 1.5. The left three bands give the resistor value, while the fourth band gives the resistance tolerance. The resistance is given by the formula:

$$R = ab \times 10^c (+/- tol\%)$$

where the *a* band is the value of the tens digit, the *b* band is the value of the ones digit, the *c* band is the base-10 exponent power value, and the *tol* band gives the tolerance or expected percentage variation in the resistor value. Table 1.1 gives these values.

Table 1.1

Resistor bands color code

a, b, c	Value	Tol Band	Value
Black	0	Silver	10%
Brown	1	Gold	5%
Red	2	Brown	1%
Orange	3	Red	2%
Yellow	4	Green	0.5%
Green	5	Blue	0.25%
Blue	6	Violet	0.1%
Violet	7	Gray	0.05%
Gray	8		
White	9		

As an example, a resistor whose bands are colored green, brown, red, and silver, respectively, has a resistance of 5.1 k +/−10% ohms.

CAPACITOR VALUES

The **capacitance of a capacitor** can be indicated in three different ways. For large capacitors, the capacitance is directly printed on the package. For small-sized capacitors, a numerical code or a color code similar to that used for resistors is used. The numerical code has three digits followed by a letter. The first two digits give the capacitance value in picofarad (pF), while the third digit is a base-10 exponent (or a multiplier of the capacitance value). The letter gives the capacitance tolerance in percentage terms. For example, a capacitor with the code 102J printed on it means it has a capacitance of 10×10^2 or 1000 pF, and the letter J means that the tolerance is 5%. Note that most small capacitors are **unpolarized**, while many large capacitors are **polarized** (have positive and negative terminals). In addition to the capacitance of the capacitor, capacitors have a voltage rating (for example, 25 V) that should not be exceeded.

Diodes and LEDs

After you complete this lab, you should be able to:

- Explain the operation of regular diodes, Zener diodes, photodiodes, and LEDs
- Know how to use these components in mechatronic applications

For this lab you need:

- 1 diode (1N4001)
- 1 Zener diode (1N5231B – 5.1V)
- 1 LED
- Resistors: 82 Ω, 100 Ω, 1 kΩ, 10 kΩ, 100 kΩ, and 1 MΩ
- BPW34 photodiode
- LM741 op-amp
- DC-DC converter (Murata NMV1215SC)

LAB BACKGROUND/INFORMATION:

Diodes and LEDs are solid-state devices. A **diode** is a directional element that is used to allow current to flow in one direction. The electrical symbol of a diode is shown in Figure 3.1a, while a schematic of a real diode is shown in Figure 3.1b. Note that when the diode is forward biased or the anode voltage is positive with respect to the cathode, current flows in the diode.

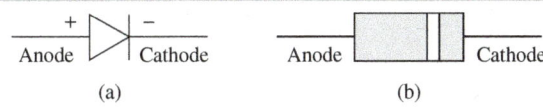

Figure 3.1

(a) Electrical symbol and (b) schematic of a real diode

A **Zener** diode is a special type of diode, and its symbol is shown in Figure 3.2. A Zener diode behaves like a normal diode when it is forward biased, but it can conduct current without destroying itself when the reverse-biased voltage exceeds the breakdown voltage (V_R). The breakdown voltage or Zener voltage (V_Z) can be smaller than that for a normal diode. Common low Zener voltages include 2.7, 3.0, 3.6, 3.9, 4.7, 5.1, 5.6, and 6.2 V, but Zener voltages of 20, 51, 100, and even 200 V are available.

An **LED** (light-emitting diode) emits light when forward biased. It is typically encased in a colored plastic casing. An advantage of an LED compared to other light

Figure 3.2

Zener diode symbol

sources is that it takes only a few milliamps (mA) to light the diode. It also can be powered by a digital power supply (5 VDC), since the voltage drop across the LED when it is on is about 2 V. A typical LED is shown in Figure 3.3. **Note that the anode or the positive terminal is the one that has a longer lead**.

Another form of a diode is the **photodiode**. A photodiode (see Figure 3.4 for a symbol) behaves like an LED but in an opposite fashion. The amount of current that the photodiode passes is proportional to the amount of light it receives, and the current flows from the cathode to the anode (reversed biased). Photodiodes commonly are used as light sensors.

Figure 3.3	
An LED	

Figure 3.4	
Symbol of photodiode	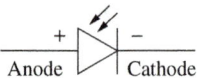

Table 6.1

(*Continued*)

Device	Symbol	Logic Function Expression	Truth Table		
OR gate	$^A_B \!\!\supset\!\!- C$	$C = A + B$	**A** **B** **C** 0 0 0 1 0 1 0 1 1 1 1 1		
NOR gate	$^A_B \!\!\supset\!\!\circ- C$	$C = \overline{A + B}$	**A** **B** **C** 0 0 1 1 0 0 0 1 0 1 1 0		
XOR gate	$^A_B \!\!\supset\!\!- C$	$C = A \oplus B$	**A** **B** **C** 0 0 0 1 0 1 0 1 1 1 1 0		
Buffer	$A -\!\!\triangleright\!- C$	$C = A$	**A** **C** 0 0 1 1		
Inverter	$A -\!\!\triangleright\!\circ- C$	$C = \overline{A}$	**A** **C** 0 1 1 0		

bar above a logic variable means the inverse of that variable. A digital logic circuit consists of a combination of basic logic devices wired together so that the output of the circuit implements a desired logical expression.

Unlike a combinational logic circuit, a sequential logic circuit output is dependent on the history of the input. A basic element of sequential logic circuits is the **flip-flop**, which is a sequential logic device that can store and switch between two binary states. The set-reset (SR) flip-flop has the symbol shown in Figure 6.1. It has two inputs, called S and R, and two outputs, called Q and complementary $\overline{Q}$.

The **555 timer chip** (such as the NE555 8-pin chip from Texas Instruments) is an integrated circuit that can produce a variety of clock signals. The pin layout for this chip is shown in Figure 6.2. The NE555 timer chip has several modes of operation. These include the **monostable** mode or fixed-pulse generation mode, and the **astable** mode or self-generating periodic signal mode. In the monostable mode, the pulse properties are controlled by one external resistor and one capacitor. In the astable mode, two external resistors and one capacitor control the duty cycle and the frequency of the timing signal.

Figure 6.1

SR flip-flop

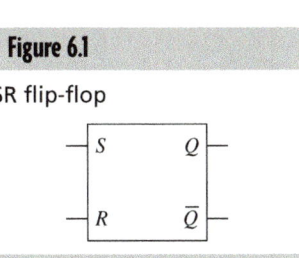

Figure 6.2

555-timer

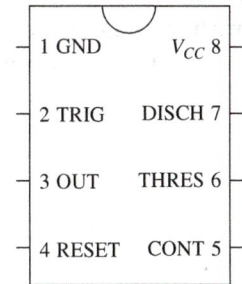

For astable operation, the on-time period (T_H) and the off-time period (T_L) are given by the following equations:

$$T_H = 0.693(R_A + R_B)C \qquad \text{(6.1)}$$

$$T_L = 0.693\, R_B C \qquad \text{(6.2)}$$

LAB 6 EXERCISES:

Name(s): _____

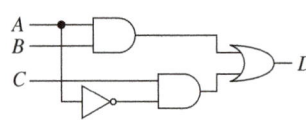

Figure E6.1

6.1 Construct circuit shown in Figure E6.1 that combines **AND, OR, and Inverter gates**. Consider all combinations of the inputs A, B, and C, and fill out the table shown below based on measurements of the D output.

A	B	C	D

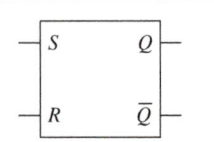

Figure E6.2

6.2 Consider the **SR flip-flop** shown in Figure E6.2. Wire up the flip-flop and fill up the table below for all combinations of S and R. Note that the SN74LS279A flip-flop has active-low inputs.

S	R	Q_t	Q_{t+1}

Figure E6.3

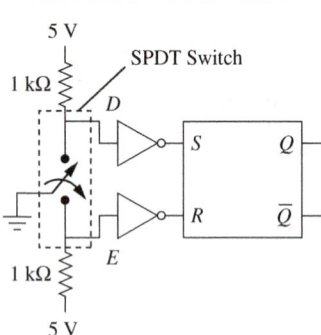

6.3 Use the SR flip-flop to construct a **switch debouncing circuit** as shown in Figure E6.3.

Use the oscilloscope to observe the output Q as the switch moves back and forth between points D and E. Describe what you observed below.

_____

_____

_____

6.4 Wire the **555 timer** as shown in Figure E6.4. Use $V_{CC} = 5$ V, $R_A = 1$ kΩ, $R_B = 1$ kΩ, and $C = 0.1$ μF. Connect the clock signal or output line (pin #3) to the oscilloscope, and measure the amplitude and the ON and OFF times of the clock signal.

Amplitude _____

ON time _____

OFF time _____

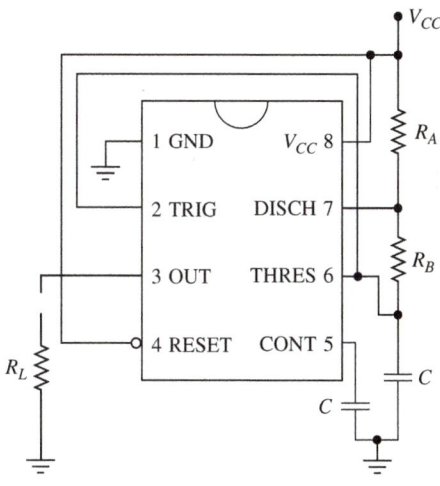

Do the measured ON and OFF times agree with the expected values? Explain.

_____

_____

_____

_____

6.5* Build a circuit that **simulates the operation of a security system** that uses two digital input signals to represent digital proximity sensors. Wire the output from these inputs such that an LED is turned ON if one input was set high, but a buzzer also is turned ON if both inputs were set high. The LED is turned OFF when neither input is high, but the buzzer is shut OFF only when neither input is high and the user has hit a reset switch. Use logic gates and a flip-flop in the design of the circuit. Draw the circuit that you designed below and demonstrate its operation to your instructor. Note that you can use two proximity sensors in this circuit instead of the digital input signals (see Lab 12 list of components).

> **Troubleshooting Tip:** Select proper resistor values so the LED and buzzer can properly operate.

QUESTIONS

6.1 What circuit family does the 74LS08 chip belong to? What type of output does it have?

6.2 How is a flip-flop different from a combinational logic gate?

6.3 List several applications where the 555 timer can be used.

6.4 For what purpose would one use two inverter gates in series?

PIC MCU—Basic

LAB BACKGROUND/INFORMATION:

A **microcontroller** is a single-chip device that includes a microprocessor, memory, and interface devices. This lab exercise demonstrates the **basics of programming and using a PIC MCU**. All of the MCU exercises in this book assume the use of the PIC16F690 MCU manufactured by Microchip, Inc., but the exercises can be done using any Microchip PIC16 or PIC18 MCU with minimal changes. Figure 7.1 shows

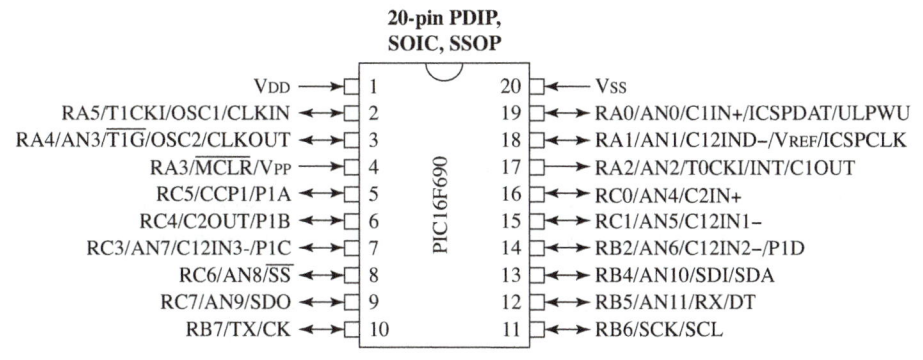

20-pin PDIP, SOIC, SSOP

	PIC16F690	
VDD →□ 1		20 □← VSS
RA5/T1CKI/OSC1/CLKIN ←→□ 2		19 □←→ RA0/AN0/C1IN+/ICSPDAT/ULPWU
RA4/AN3/T1G/OSC2/CLKOUT ←→□ 3		18 □←→ RA1/AN1/C12IND−/VREF/ICSPCLK
RA3/MCLR/VPP →□ 4		17 □→ RA2/AN2/T0CKI/INT/C1OUT
RC5/CCP1/P1A ←→□ 5		16 □←→ RC0/AN4/C2IN+
RC4/C2OUT/P1B ←→□ 6		15 □←→ RC1/AN5/C12IN1−
RC3/AN7/C12IN3-/P1C ←→□ 7		14 □←→ RB2/AN6/C12IN2−/P1D
RC6/AN8/SS ←→□ 8		13 □←→ RB4/AN10/SDI/SDA
RC7/AN9/SDO ←→□ 9		12 □←→ RB5/AN11/RX/DT
RB7/TX/CK ←→□ 10		11 □←→ RB6/SCK/SCL

Figure 7.1

PIC16F690 MCU

(Reprinted with the permission of Microchip Technology Incorporated.)

the pin layout for the PIC16F690 chip. Due to the fact that this MCU has only twenty pins but supports many interface functions, many of the pins are designed for more than one function.

In using a microcontroller, one needs to develop and download a program to the MCU that has instructions for the MCU to execute. The program can be developed using a high-level programming language (such as C or Visual Basic (VB)) but also can be developed using assembly language. This lab makes use of a **C-language compiler developed by CCS, Inc.** This compiler will be referred to as the PIC-C compiler. (Note that another C compiler can be used to perform the MCU-related exercises in this lab book, but there could be slight differences when referring to pin names in the code or in the format and operation of the compiler-provided functions). The compiled code is in binary form (or hex code). The hex code needs to be downloaded to the microcontroller and stored in the nonvolatile program memory section of the MCU before the program can execute.

A **C-program** is made up of the following elements in a file (type .c)

- Comments

- Pre-processor directives

- Data/variable definitions

- Function definitions

A **C program must have a function named** *main( )*, which is the starting point for program execution. The program could have other functions that are called from *main( )* or from other functions. Each statement in the program must end with a semicolon. Braces ({ }) are used to define code sections. The structure of a typical C program is shown in Figure 7.2. A screen shot of the Integrated Development Environment (IDE) for the PIC-C compiler is shown in Figure 7.3. To compile a program, the user simply clicks on the *Compile* menu item in the *Compile* tab. The PIC-C compiler uses software **fuses** to set the operating environment for the particular MCU.

Figure 7.2

Structure of a typical C program

```
Comment Line              // File: Digital_Input.c
                          #include <18F8722.h>
Preprocessor directives   #fuses HS, NOMCLR, NOWDT
                          #use delay (clock=10000000)
Data                      int8 value;
                          void main()
                          {
                             while (2 > 1) // Start infinite loop
Main Routine                 {
                                output_high(PIN_D0);//Status LED
                                ....
                             }
                          }
```

The process of transferring a compiled binary code to the MCU is called 'programming' a chip. Microchip has currently two programmers called the **PICkit™ 2** and the **PICkit™ 3** microcontroller programmers. They are low-cost development programmers that can be conveniently used to program many MCU chips. Both the PICkit 2 and the PICkit 3 programmers (see Figure 7.4) connect to a PC through a USB cable, and using the software that comes with them, one can download programs to the PIC MCU. Both programmers conveniently plug into a number of development boards (examples of such boards are shown in Figures 7.5 and 7.6). The **low pin-count development board** has four LEDs, a single-turn potentiometer, and a switch.

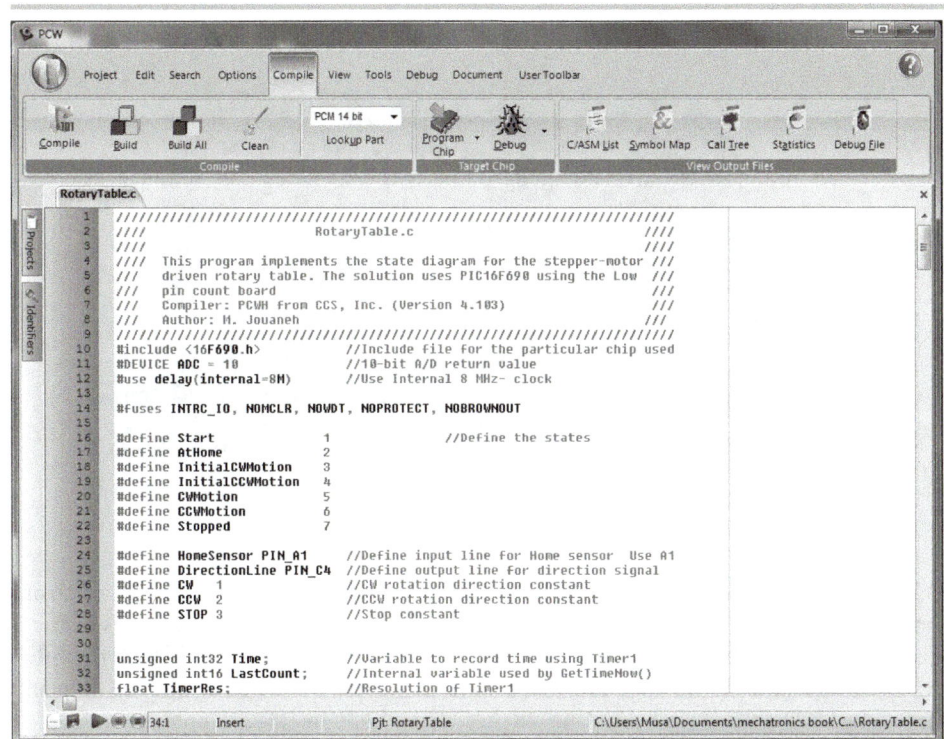

Figure 7.3

IDE for PIC-C compiler

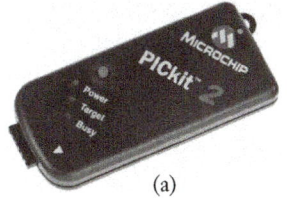
(a)

(b)

Figure 7.4

(a) PICkit™ 2 and
(b) PICkit™ 3 programmers
(Reprinted with the permission
of Microchip Technology
Incorporated.)

SW1
switch

Rotary
pot

LED1

Figure 7.5

Microchip low pin-count
development board
(Jouaneh, University of Rhode Island.)

Figure 7.6

Microchip PICDEM PIC18
development board
(Jouaneh, University of Rhode Island.)

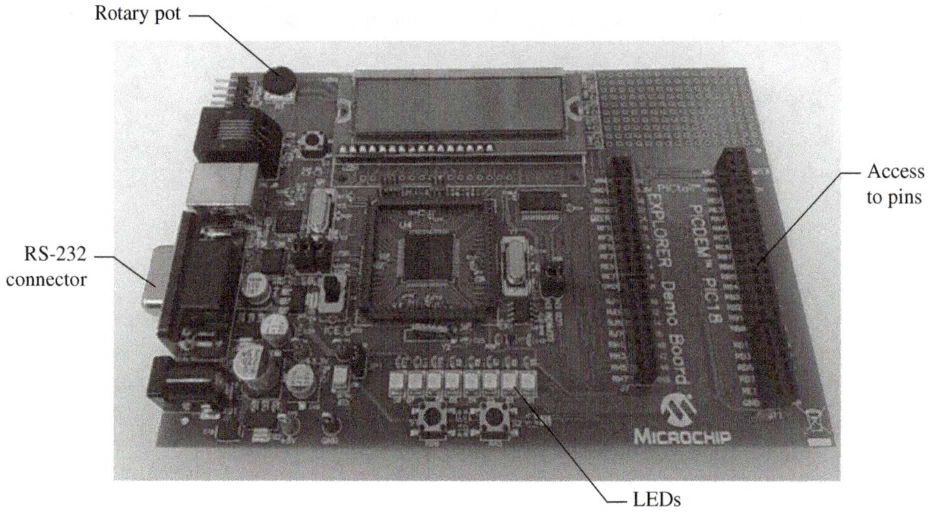

A development board can be used to run code without the need to build a custom board to house the PIC MCU. Microchip manufactures many different types of these development boards. Another development board is the **PICDEM™ PIC18 exploration board** (see Figure 7.6). This development board uses the PIC18F8722 chip and has eight LEDs, three push-button switches, and a built-in RS-232 connecter.

The PICkit 2 programmer can be used from the MPLAB IDE (an integrated editor and compiler for PIC MCUs that is provided by Microchip) or through the use of a separate stand-alone program called the PICkit 2 program, which Microchip also provides. There is no stand-alone program for PICkit 3, so PICkit 3 has to be called from the MPLAB IDE. The PICkit 3 also can be used as a debugger to step and examine a code as it executes.

The **steps needed to download a code to a PIC MCU using the PICkit 2 or PICkit 3 programmers** are listed here.

Step 1. **Develop** program code in C.

Step 2. **Compile** the developed code to create the binary (or hex) code that resides on the MCU. The compiled code is called the hex code since the compiled instructions are written in hexadecimal notation.

Step 3. **Connect** the development board on which the MCU chip is mounted to the PICkit 2 or PICkit 3 using the 6-pin connecter on the programmer.

Step 4. **Plug** the USB cable of the PICkit 2 or PICkit 3 into the PC which has the compiled code.

Step 5. **Call up** the PICkit 2 program or MPLAB IDE (if using PICkit 3). Note that if you called the PICkit 2 program without plugging first the PICkit 2 programmer to the PC, you will get an error message.

Step 6. **Open** the hex file to be downloaded (use *File\Import Hex*).

Step 7. **Download** the program to the MCU. If successful, you will get the message 'Programming Successful.'

Note that the PICkit 2 programmer program can be used to program an MCU without the need to have power supplied to the board. The PICkit 2 programmer program can also provide power to the connection target (provided that the power drawn from the development board is minimal). This is done by checking the *On* check box in the *VDD PICkit 2* group area (see Figure 7.7) and adjusting the VDD voltage value (normally set to 5.0 V).

Figure 7.7

PICkit 2 interface
(Reprinted with the permission
of Microchip Technology
Incorporated.)

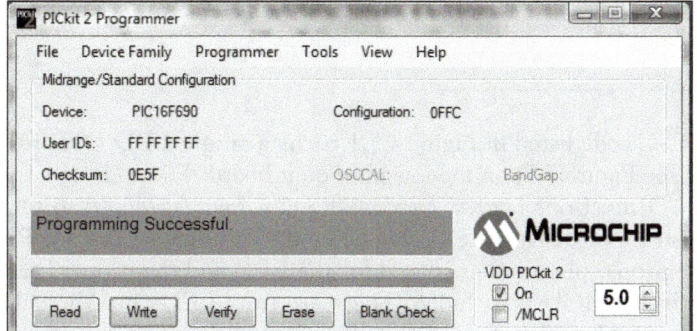

For I/O, the PIC-C compiler provides functions that can affect a single bit or the entire port. The **single pin or bit functions** are

output_high(pin)	//Set the selected pin to high
output_low(pin)	//Set the selected pin to low
output_pin(pin, value)	//Send a specified value (0/1) to selected pin
input(pin)	//Returns the state of the selected pin

and the **port functions** are

output_x(value)	//Send an entire byte to port x (x = a, b, c, d, . . .)
input_x()	//Read an 8-bit integer representing the port input value

The compiler has directives to specify the type of input/output. In the *STANDARD_IO* (default method) mode (*#use standard_io(port_name)*), the compiler automatically generates code to make an I/O pin either input or output every time it is used. The tristate register is automatically updated in this mode. In the *FAST_IO* mode (*#use fast_io(port_name)*), the compiler will perform I/O without programming of the direction register. The user has to set the direction by calling the set tristate register function (*set_tris_x()*).

LAB 7 EXERCISES:

Name(s): _____

7.1 The PIC-C code listed in Figure E7.1 **turns a single LED ON and OFF** (LED1, see Figure 7.5) on the low pin-count board. The code uses the single-pin digital I/O functions (*output_high(pin#)* and *output_low(pin#)*) to turn the LED ON or OFF. Note that the code can be used with any board or any PIC MCU, as long the proper pin number is used for the LED and the proper header file for the particular MCU is included. There are **four LEDs** on Microchip low pin-count board. These LEDs are internally connected to pins 0 through 3 of **digital Port C**. The code uses the internal 8 MHz clock on the PIC microcontroller as the clock source. Use the provided code to gain familiarity with downloading a program to the PIC microcontroller. Compile your code, and download the program to the chip using the instructions provided in the previous section.

Your program should run automatically, and you should see LED1 turn ON and OFF every two seconds.

Figure E7.1

```
///////////////////////////////////////////////////////////////////////////////////////
///              BitOnOff.c
///  This program turn on/off bit 0 of Port C
///
///   Compiler: PCWH from CCS, Inc. (Version 4.103)
///////////////////////////////////////////////////////////////////////////////////////
#include <16F690.h>                        // Include file for the particular chip used
#use delay(internal=8M)                    // Use Internal 8 MHz- clock

#fuses INTRC_IO, NOMCLR                     // Set clock mode to internal oscillator with
                                           // no clkout. Master clear pin is used for I/O
void main(void)                            // Main function
{
   while (2 > 1)                           // Start infinite loop
   {
   output_high(PIN_C0);                    // Set the selected pin to high
   delay_ms(1000);                         // Set a time delay of 1000 ms
   output_low(PIN_C0);                     // Set the selected pin to low
   delay_ms(1000);                         // Set a time delay of 1000 ms
   }
}
```

Revise the code to make any **two LEDs on the board** turn ON and OFF in each cycle (instead of a single LED). Compile and download your program to the MCU. Print a copy of the revised code and attach it here.

_____

7.2 Revise the code in Exercise 7.1 to perform the following.

a. Add code to read **digital input** from switch SW1 (see Figure 7.5) on the low pin-count board in each cycle of the code. Switch SW1 is internally connected to pin A3 on the board and gives a digital low when pressed down. Use the *input(pin)* function to read the switch value.

b. When switch SW1 is not pressed down, the program still turns ON and OFF LED1 as before. While switch SW1 is pressed down, make the code turn on LEDs 1 and 3 for half a second while LEDs 2 and 4 are OFF. In the next half-second, alternate the pattern to make LEDs 1 and 3 OFF while LEDs 2 and 4 are ON.

Test your code to make sure it works as described above. Print a copy of your code and attach it here.

Does the LED pattern change immediately when you press/release switch SW1?

_____

_____

_____

7.3 Using the breadboard, build a simple circuit that uses the transistor and the 1 kΩ resistor to **turn ON and OFF a small DC motor** from code running on the PIC MCU. The base input of the transistor should be connected to pin C0 on the low pin-count board. Place a diode (called a **flyback** diode in this application) across the motor leads to protect the transistor with the diode cathode side connected to V_{CC}. Check the voltages at all nodes using the multimeter. Draw a copy of the needed circuit below. Use the same code that was used for Exercise 7.1, but change the ON and OFF times to 5 seconds each.

Explain what you observe when you run the program.

_____

_____

Now change the ON and OFF times to 0.5 s, and run the code again. Explain what you observe when you run the program now.

_____

_____

_____

7.4* Write a program that **scrolls through all the digits (0 through 9) in the 7-segment display**. Place the 7-segment display on the breadboard, and use the binary coded decimal (BCD) decoder chip as an interface between the PIC16F690 and the 7-segment display. Use four outputs lines from the PIC to transmit the BCD pattern of each digit to the decoder chip. A schematic of the 7-segment digital display is shown in Figure E7.4. Refer to the data sheets for the display and the decoder chips. Attach a printout of the code you developed and demonstrate the operation of this program/circuit to your instructor.

Figure E7.4

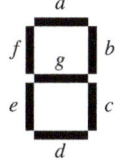

> **Troubleshooting Tip:** If the digits do not show up properly, check the data sheet for the proper wiring of the LE pin.

QUESTIONS

7.1 How can a program be made to run on an MCU?

7.2 What is a hex code?

7.3 In what part of the MCU are variable declarations stored?

7.4 What is the advantage of using an external clock source with an MCU?

PIC MCU—A/D and PWM

After you complete this lab, you should be able to:

- Explain the operation of the A/D converter and the PWM module on a PIC MCU
- Write code for using the A/D converter on the PIC MCU
- Write code for using the PWM module on the PIC MCU

For this lab you need:

- TIP29 transistor
- 1 kΩ resistor
- Diode (1N4001)
- Small 5 VDC motor
- Hitec HS-311 hobby servo motor

| LAB BACKGROUND/INFORMATION:

In this laboratory exercise, we will explore additional features on the PIC MCU. These include the use of an A/D convertor and a pulse-width modulated output. The **PIC MCU A/D converter** converts analog voltages to digital numbers. The analog voltage range is 0 to VDD (0 to 2.0 to 5.5 V or smaller if V_{ref+} is set). The resolution is 8, 10, or 12 bits. On the PIC16F690, the resolution is 10 bits, but the output can be mapped into 8 bits. The bit resolution affects the **voltage resolution** or increment of the A/D converter which is defined as

$$\text{Voltage resolution} = \text{range}/2^n \tag{8.1}$$

where n is the bit resolution of the A/D convertor.

The **digital output** of an A/D converter subjected to an input voltage V_{in} is given by

$$\text{Digital output} = \text{ceiling} \left((V_{in} - V_{ref-} - \text{voltage resolution})/\text{voltage resolution} \right) \tag{8.2}$$

where V_{ref-} is the negative reference voltage (0 V in most cases).

To use the A/D converter, the PIC-C compiler provides several functions for this purpose. **First, the user needs to call the function *setup_adc( )*, which sets up the clock source for the A/D converter**. The user can select either a sub-frequency of the oscillator frequency (such as $F_{OSC}/16$) or the A/D dedicated internal oscillator (F_{RC}) as the clock source. **Next, one needs to select which pins on the MCU will be used for the A/D conversion and the voltage reference to use when computing**

the A/D value. This is done by calling the function *setup_adc_ports( )* function. Using this function, all, some, or none of the channels can be set to perform A/D. In addition, this function can be used to specify the channel number to be used as the analog reference voltage if a voltage reference other than VDD is used. The above two functions need only to be called once.

To read a particular A/D channel, the channel needs to be selected first by calling the *set_adc_channel( )* before the *read_adc( )* function is called. If the same A/D channel is read each time, then only the *read_adc( )* function has to be called after the *set_adc_channel( )* function was called once. The bit resolution of the A/D conversion is set by including the directive *#DEVICE ADC = Num_of_bits* at the top of the C-language file.

None of the MCUs in the PIC16 or PIC18 families have D/A capability, but many PIC MCUs have a built-in module to generate **pulse-width modulated (PWM) output** (such as pins 5, 6, 7, or 14 on the PIC16F690 (see Figure 8.1)). The PWM output can be used to conveniently drive H-bridge drivers and digital amplifiers. The user can vary the frequency and duty cycle of the fixed voltage square-wave output signal. The PWM output mode is one of the three modes of operation of the **C**apture/**C**ompare/**P**WM (or CCP) module or the **E**nhanced CPP (ECCP) module on the MCU. The CPP module is configured for PWM operation in the PIC-C compiler by calling the function ***setup_ccp1(CCP_PWM)***, which configures, for example, the P1A channel to operate in PWM mode.

Figure 8.1	**20-pin PDIP, SOIC, SSOP**
PIC16F690 MCU (Reprinted with the permission of Microchip Technology Incorporated.)	

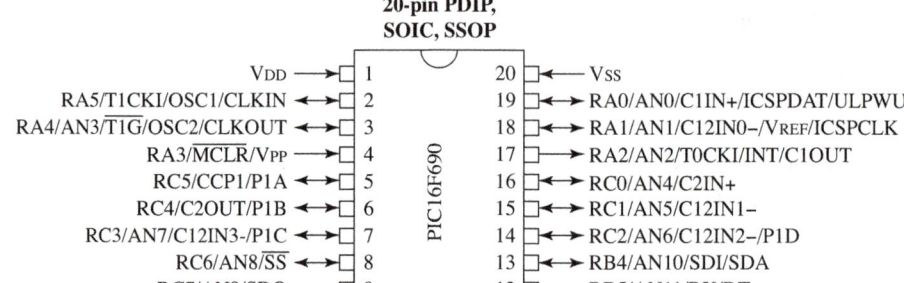

Timer 2 is used to control the frequency of the PWM signal. **The frequency of the PWM signal** is given by

(8.3)

$$\text{PWM}_{\text{freq}} = (F_{\text{OSC}}/4)/((1 + \text{PR2}) \times t2\text{pres})$$

where F_{OSC} is the oscillator frequency, $t2_{\text{pres}}$ is the prescaler factor for Timer2, and *PR2* is the Timer2 period register value (0 to 255). For example, at a clock frequency of 8 MHz, a $t2_{\text{pres}}$ value of 4 and a Timer2 period register value of 255, the PWM frequency will be 1.953 kHz.

Using the PIC-C compiler, **the command to set up Timer2** to obtain the given frequency is

setup_timer_2(T2_DIV_BY_4, 255,1)

where the first argument is the prescale factor, the second argument is the Timer2 register value, and the third argument is the Timer2 postscaler value. Note that Timer2 postscaler is not used in the determination of the PWM frequency. **The duty cycle ratio is given by**

(8.4)

$$\text{Duty\_cycle ratio} = \text{value}/(4 \times (1 + \text{PR2}))$$

where *value* is the parameter of the *set_pwm1_duty( )* function in the PIC-C compiler. For a 50% duty cycle and a PR2 value of 255, *value* is 512 in Equation (8.4).

Note that while the PIC16F690 has four PWM output channels, they all run from a single PWM generator. This means that the user cannot independently control the frequency and duty cycle of each of these channels. The user can, however, control which channel is used for PWM output (default channel is P1A on PIC16F690). This is done through the Pulse Steering Control (PSTRCON) register (see data sheet for details). PIC MCUs with ECCP module (such as PIC16F690) can also be configured for full or half H-Bridge PWM control (see data sheet for details).

LAB 8 EXERCISES:

Name(s): _____

8.1 The PIC-C program shown in Figure E8.1 implements an infinite loop which **reads the input from the analog-to-digital converter channel 0** and turns ON LED1 if the read value is less or equal to 512 (OFF otherwise). A/D channel 0 is internally connected to the rotary pot on Microchip low pin-count development board. Note the three lines of code (*setup_adc( )*, *setup_adc_ports( )*, and *set_adc_channel( )*) to set and select port RA0 on the PIC16F690 to operate as A/D channel 0. Compile and download this code to the MCU. Rotate the pot by hand and verify the operation of the program.

Figure E8.1

```
////////////////////////////////////////////////////////////////////////////////////
///                  Analog_Input_LED.c
///
///   This program set on/off LED on pin C0 based the value read
///   from A/D channel 0 (RA0)
///
///   Compiler: PCWH from CCS, Inc. (Version 4.103)
////////////////////////////////////////////////////////////////////////////////////
#include <16F690.h>                        // Constant definitions for chip used
#DEVICE ADC = 10                           // 10-bit A/D return value
#fuses  INTRC_IO, NOMCLR                    // Set clock mode to internal oscillator with
                                           // no clkout. Master clear pin is used for I/O
#use delay (INTERNAL=8M)                    // Use Internal 8 MHz clock
void main()
{
int16 addata;
setup_adc(ADC_CLOCK_DIV_16);               // Set A/D clock frequency as FOSC/16
setup_adc_ports(sAN0);                     // Set channel 0 for A/D
set_adc_channel(0);                        // Select channel 0
while (2 > 1)                               // Start infinite loop
  {
  addata = read_adc();                     // Read selected A/D channel
  if (addata <= 512)
    {
      output_high(PIN_C0);                 // Set port C0 to high
    }
  else
    {
    output_low(PIN_C0);                    // Set port C0 to low
    }
    delay_ms(1000);                        // Wait 1000 ms
  }
}
```

8.2 Modify the code given in Exercise 8.1 to **map the A/D reading from channel 0 to the four LED outputs on the development board.** Divide the A/D output of 0-1023 into 16 intervals (i.e., 0 to 63, 64 to 127, 128 to 191, etc.). Based on which interval the output of the A/D falls into, turn on the appropriate number of LEDs to indicate the interval number to the user using a 4-bit binary number (see the first two columns in the table below). For example, if the A/D output is within the interval of 256 to 319, then the binary number 4 should be indicated by the LEDs. Consider LED1 to be the least significant bit of the 4-bit binary number to be displayed. Compile and download the code into the PIC MCU. Test the code by slowly rotating the rotary pot over its operating range, and all of the binary numbers 0 through 15 should be displayed by the LEDs. Attach a printout of the code you developed.

A/D Channel Zero Output Value	Number Displayed by LEDs	PWM Duty Cycle (%)
0–63	0	0
64–127	1	6.25
128–191	2	12.5
192–255	3	18.75
256–319	4	25
...	...	...
896–959	14	87.5
960–1023	15	93.75

Hint: Consider using a *port* output function to make the code more efficient.

8.3 Modify the code you developed in Exercise 8.2 so you can **control the speed of a small DC motor** using the PIC16F690 MCU. The modified code should still monitor the analog input from A/D channel zero, but it will use the interval that the A/D output falls in to set the PWM duty cycle for control of the motor speed. The A/D output values should be mapped into sixteen intervals, and for each interval, a different PWM duty cycle should be sent as well as the number displayed by the LEDs. The specifics are listed in the table shown in Exercise 8.2. The PWM output from the micro-controller should be connected to the base input of the TIP29 transistor through a resistor. Place a flyback diode across the motor leads. Build the interface circuit on the breadboard. Explain below what you observed when you ran the code. Attach a printout of the revised code.

Hint: Vary the PWM frequency and observe its effect on motor response.

Troubleshooting Tip: Use *int16* variable type for the duty cycle parameter of the *set_pwm1_duty( )* function.

8.4* Develop a program to **turn the four LEDs on the board ON and OFF in a particular pattern** (one ON and the next OFF, two ON and the next two OFF, . . .). You can create any pattern you like. Use the A/D reading from the pot input (Channel RA0) to vary the timing speed of the pattern. Turning the pot CW (as seen from above) should cause the pattern to turn ON and OFF more rapidly. Use the *delay_ms( )* function to create the timing delay. Print a copy of the code you developed and demonstrate the operation of the program to your instructor.

Figure E8.5

A standard size servo
(Courtesy of Hitec RCD USA, Poway, CA.)

8.5* Use the **PWM feature on the microcontroller to drive the Hitec HS-311 hobby servo motor** shown in Figure E8.5. The control signal for this servo motor is a PWM signal (5 V) at a frequency of 20 to 60 Hz. The pulse width of the control signal, which ranges from 0.7 to 2.3 ms, controls the position of the hobby servo. At 0.7 ms, the servo is at one extreme of its motion range, while at 2.3 ms, it is at the other extreme. At a 1.5-ms pulse width, the servo is in the center or mid-position. This servo has a motion range of ± 90°. Use the rotary pot to adjust the pulse width (or duty cycle) setting of the control signal. When the rotary pot is at one extreme of its travel, the pulse-width setting should be 0.7 ms. Turning the pot clockwise from that position should increase the pulse-width setting. When the rotary pot is at the other extreme of its motion, the pulse-width setting should be 2.3 ms. Implement an infinite *Do-loop* to read the desired pulse width and to adjust the duty cycle of the PWM signal. Add a small delay (100 ms) in each run through the loop. Print a copy of the code you developed and demonstrate the operation of the program to your instructor.

QUESTIONS

8.1 How can the A/D converter voltage resolution be improved?

8.2 What causes the motor in Exercise 8.3 to not move at low duty cycles?

8.3 What is the minimum PWM frequency that can be obtained using an 8 MHz oscillator frequency?

PIC MCU—Serial Interfacing

After you complete this lab, you should be able to:

- Explain the different serial interfaces available on a PIC MCU
- Write code for RS-232, SPI, and I^{2}C interfacing

For this lab you need:

- MAX233
- DS1631
- Two 1 KΩ resistors
- MCP42010-I/P
- 25LC256
- Serial cable or USB-to-serial converter (Sewell Direct SW-1301)

LAB BACKGROUND/INFORMATION:

This laboratory explores serial interfacing in PIC MCUs. A **serial port** is an input/output device that takes data in a parallel form and transmits it in a serial fashion. Terminals, modems, mice, and keyboards are examples of devices that connect to a PC/MCU through a serial port. In a serial port, the data is transmitted one bit at a time rather than simultaneously. The data to be transmitted is first broken up into packets, each packet is made up of a number of bits, and then the bits in these packets are transmitted sequentially (thus, the name serial port). In a PIC MCU, the serial port is part of the built-in **USART** (Universal Synchronous Asynchronous Receiver Transmitter) module or a variation of it, such as the **EUSART** (Enhanced USART) or the **AUSART** (Addressable USART) modules.

Note that while a USART provides the mechanism for converting parallel data into serial data and vice versa, additional hardware is needed to wire a PIC MCU to a PC for serial RS-232 communication. This is needed since the PIC does not supply the voltages needed (up to $+/- 15$ V) for serial communication with a PC. A commonly used device is the MAX232 or the MAX233 interface chip. The connection diagram using a MAX233 chip is shown in Figure 9.1. Note that the MAX233 chip requires no external capacitors, while the MAX232 requires several capacitors.

The common method of using a USART on a PIC MCU is to implement full-duplex asynchronous serial communication using the RS-232 protocol. **The CSS compiler has several functions for serial communication**. These include the *getc( )*, *gets( )*, *putc( )*, *printf( )*, and *kbhit( )* functions. The **getc( )** gets a single character from the buffer. The *getc( )* function is blocking, and thus, one should use first the **kbhit( )** to determine if a character is available in the buffer before calling the *getc( )* function.

Figure 9.1

Wiring between a PIC16F690
and a MAX233

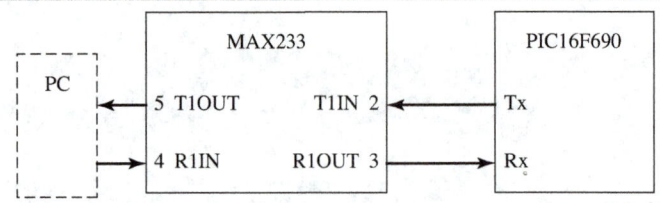

The *gets( )* gets a string from the buffer until a carriage return is encountered. The *putc( )* places a character in the output buffer, while *printf( )* sends a formatted string to the buffer.

To provide higher communication speeds, many PIC chips have a built-in synchronous serial port interface module. The PIC16F690 supports both the **Serial Peripheral Interface (SPI)** and the **Inter-Integrated Circuit (I²C™) interface**. The SPI operates in full duplex and at speeds of 1 Mbps or higher. It is simple to implement (needing only four wires) and uses the concept of master/slave. The I²C interface is a synchronous serial communication protocol that was developed by Philips Semiconductor. The I²C or the Inter-Integrated Circuit (I²C™) interface uses just two wires: one for data transmission and the other for the clock signal for the interface between two devices.

LAB 9 EXERCISES:

Name(s): _____

9.1 **Build a circuit to interface the PIC16F690 with the serial port on a PC**.
Use the MAX233 chip and build the interface circuit shown in Figure E9.1. If
you are using a demonstration board with a built-in RS-232 connector, then
there is no need for you to build this circuit. Note that the MAX233 chip performs the same job as the MAX232 chip, but it does not need any capacitors.

Figure E9.1

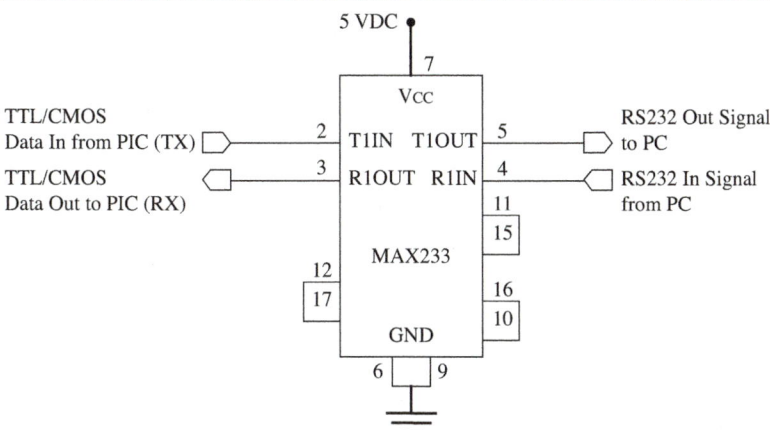

> **Hint:** If the PC you are using does not have a built-in RS-232 port, you need to use
> a USB-to-serial convertor cable. In this case, you also need to load a driver on the PC
> to do the communication. The driver is available from the USB-to-serial cable
> manufacturer.

9.2 Use a **terminal program** such as *PuTTY (available from http://www.chiark.greenend.
org.uk/~sgtatham/putty/)* **to communicate with the PIC16F690**. Program the
MCU with the code shown in Figure E9.2 to demonstrate basic serial interfacing. In this code, the *main()* routine has an infinite loop which continuously
scans the serial input buffer. If the character read from the port is 'a', then the
program sends the character '1' to the terminal program.

Figure E9.2

```
///////////////////////////////////////////////////////////////////////////
///                    Serial_In_Out_Lab.c
///
///     Program that demonstrates RS-232 communication
///     Compiler: PCWH from CCS, Inc. (Version 4.103)
///////////////////////////////////////////////////////////////////////////
#include <16F690.h>
#fuses INTRC_IO, NOMCLR, NOWDT
#use delay (internal=8M)
#use rs232(baud=38400, BITS = 8, PARITY = N, xmit=PIN_B7, rcv=PIN_B5)
char ReadSer();                        // Function prototype statement
void WriteSer(char);                   // Function prototype statement

void main()
{
char c;
while (2 > 1) // Start infinite loop
   {
       c  = ReadSer();                 // Read the serial  port
       if (c == 'a')
         {
         WriteSer('1');                // If the read char is 'a', then send the char '1'
         }
       }
}
char ReadSer()                         // A function that reads a character from the serial port
{
  if (kbhit() == 1)                    // The kbhit() function is used because the getc() function is blocking
     return(getc());                   // if there is no character in the serial buffer
  else
     return('0');
}
void WriteSer(char d)                  // A function that sends a char to the serial port
{
  printf("%c\n\r",d);                  // Send new line (\n) and carriage return (\r) characters
}
```

Next, **modify the provided code** to enable the terminal program to send command strings (not just individual characters) to the PIC MCU to enable the MCU to turn ON or OFF one or more LEDs based on these commands. Use the *gets()* function in your modification. At a minimum, modify your code to support the following two commands: *ON* (which should cause LED1 or any other LED to turn on) and *OFF* (which should cause the same LED to turn off). Attach a printout of your modified code.

> **Troubleshooting Tip:** Make sure that the communication parameters (such as baud rate and parity) have the same setting in the PC terminal program and in the PIC MCU code.

9.3 The 25LC256 is a **serial EEPROM** with 32 Kbyte capacity **that uses the SPI interface** to read and write to the memory. The code listing shown in Figure E9.3 has three functions: one to initialize the EEPROM, another to write data to memory, and the third to read data from memory. Refer to the data sheet of this chip to understand the operation of these functions. For this exercise, do the following.

   a. Build a circuit on a breadboard to interface this EEPROM chip to the PIC16F690 MCU or another PIC MCU.

```
///////////////////////////////////////////////////////////////////////////////////
///                 Test_25LC256.C
///
///   A collection of routines for SPI communication with the 25LC256 EEPROM
 ///   Compiler: PCWH from CCS, Inc. (Version 4.103)
///////////////////////////////////////////////////////////////////////////////////

#define EEPROM_ADDRESS long int
#define READ        0x03
#define WRITE       0x02
#define WREN        0x06

void init_ext_eeprom()                               // Initialize code - called once
  {
    output_high(EEPROM_SELECT);                      // Make EEPROM_SELECT line high
    setup_spi(SPI_MASTER |SPI_XMIT_L_TO_H| SPI_CLK_DIV_16 ); // Set SPI mode
  }

void write_ext_eeprom(EEPROM_ADDRESS address, BYTE data)
  {
    output_low(EEPROM_SELECT);                       // Make EEPROM_SELECT line low
    spi_write(WREN);                                 // Send code to enable writing
    output_high(EEPROM_SELECT);                      // Make EEPROM_SELECT line high
    output_low(EEPROM_SELECT);                       // Make EEPROM_SELECT line low
    spi_write(WRITE);                                // Send code to start writing
    spi_write(address>>8);                           // Send MSB byte first
    spi_write(address);                              // Send LSB byte
    spi_write(data);                                 // Send data
    output_high(EEPROM_SELECT);                      // Make EEPROM_SELECT line high
    delay_ms(6);                                     // Delay to complete the erase and writing of data
  }

BYTE read_ext_eeprom(EEPROM_ADDRESS address)
  {
    BYTE data;
    output_low(EEPROM_SELECT);                       // Make EEPROM_SELECT line low
    spi_write(READ);                                 // Send code to start reading
    spi_write(address>>8);                           // Send MSB byte first
    spi_write(address);                              // Send LSB byte
    data=spi_read(0);                                // Read data
    output_high(EEPROM_SELECT);                      // Make EEPROM_SELECT line high
    return(data);                                    // Return data to calling function
  }
```

b. Use the provided functions to develop a program that allows the user to read and write data to this chip using the SPI interface. The program should allow the user through the use of a terminal program such as *PuTTY* to specify the type of operation to be done (read or write). For read operations, the user should supply the desired memory address, while for write operations, the user should supply the memory address as well as the data to write to the specified address.

Provide a schematic of the circuit plus a printout of the code you developed.

Hint: Use the function *gethex( )* (you will need to add the statement *#include <input.c>* to your program) to get the address or the data from the terminal program. The function *gethex( )* returns a byte in hexadecimal notation.

> **Troubleshooting Tip:** You need to specify the pin number for the *EEPROM_SELECT* line in your code. The *HOLD* line on the 25LC256 should be connected to V_{CC}.

> **Troubleshooting Tip:** The SDO line on the PIC should be connected to the SI pin on the chip, and the SDI line on the PIC should be connected to the SO pin on the chip.

9.4* The **DS1631 is a digital temperature sensor that uses I²C interfacing** and measures temperature over the range of −50 to 125°C. With reference to the data sheet of MAXIM DS1631, do the following.

  a. Build a circuit on a breadboard to interface this sensor to the PIC16F690 MCU or another PIC MCU.

  b. Develop a program to read the room air temperature measured by the sensor. Note that your program should convert the digital sensor output into a temperature in engineering units. The conversion (see details in data sheet) is dependent on the sensor temperature measurement resolution. The default resolution is 12 bits. Display the read temperature by transmitting it to a terminal program for display using RS-232 interfacing.

  c. Provide a schematic of the interface circuit you developed as well as the program to read the sensor.

> **Hint:** Structure your code to have two functions: one to initialize the DS1631 and the other to read the temperature. Set the DS1631 to operate in continuous temperature conversion mode.

> **Troubleshooting Tip:** The I²C SDA and SCL lines are open collector types and require a pull-up resistor.

9.5* With reference to the data sheet of **Microchip MCP42010-I/P (a digital potentiometer that uses SPI interfacing)**, do the following.

  a. Build a circuit on a breadboard to interface this potentiometer IC to a PIC MCU.

  b. Develop a program to set the resistance of the potentiometer. The resistance is set by sending a byte that changes the resistance in one of 256 steps. Use a multimeter to measure the value of the resistance at the potentiometer leads.

Provide a schematic of the interface circuit you developed as well as the program you developed. Note that the MCP42XXX is a 256-position digital potentiometer that has two independent channels and is available in 10 kΩ, 50 kΩ, and 100 kΩ versions. At startup, the potentiometer wiper position is at the mid-resistance value. You could combine the MCP4010 with an op-amp to create a programmable single-ended or differential amplifier.

Hint: Structure your code to have two functions: one to initialize the MCP42010-I/P and the other to write a byte to set the resistance.

Troubleshooting Tip: If you have difficulty changing the resistance of the potentiometer, it most probably may be caused by not specifying the correct mode for the SPI data transfer.

Troubleshooting Tip: The SDO line on the PIC should be connected to the SI pin on the chip, and the SDI line on the PIC should be connected to the SO pin on the chip.

QUESTIONS

9.1 What is the difference between RS-232 serial communication and SPI communication?

9.2 What is the purpose of using a MAX 232/233 interface chip?

9.3 Determine the transmission data speed for SPI communication in Exercise 9.3.

PIC MCU—Timing and Interrupts

After you complete this lab, you should be able to:

- Know how to use timers on a PIC MCU
- Explain the operation of interrupts
- Write code for handling certain interrupts

For this lab you need:

- RS-232 interface circuit between a PIC MCU and a PC or a PIC development board with a built-in RS-232 interface
- A serial cable or a USB-to-serial cable (Sewell Direct SW-1301)

| LAB BACKGROUND/INFORMATION:

Time is a very important element in software used in mechatronic applications. **Timing needs** in mechatronic applications include: the use of a timer to record the time of occurrence of an event (such as recording when the temperature of a process reaches a certain value), the use of timers to implement time delays, and the use of timers to schedule repeated execution of code segments (such as those used for monitoring and feedback control).

This laboratory explores timing and interrupts in PIC MCUs. The PIC16F690 has three **timers** called Timer0, Timer1, and Timer2. Table 10.1 lists information about these timers.

Table 10.1		Timer0	Timer1	Timer2
Timers in PIC16F690 microcontroller	Bit Size	8-bit	16-bit	8-bit
	Operate as a Counter?	Yes	Yes	No
	Programmable Prescaler?	Yes	Yes	Yes
	Prescaler Values	1:1 to 1:256	1:1 to 1:8	1:1, 1:4, 1:16
	Postscaler?	No	No	Yes
	Postscaler values	—	—	1:1 to 1:16
	Maximum Timer Interval at 8 MHz Clock (with maximum prescale)	$0.128 \times 256 = 32.768$ ms	$32.768 \times 8 = 262.1$ ms	$0.128 \times 16 = 2.048$ ms
	Timer Overflow Interrupts	Yes	Yes	Yes

Note that all PIC microcontrollers execute instructions at a rate that is ¼ of the clock rate. For example, if a microcontroller was running on an 8 MHz clock, then the instruction cycle rate is 2 MHz. The use of a programmable **prescaler** reduces the instruction-cycle rate as seen by the timer, allowing for longer timing intervals to be counted by the timer before overflow. For example, if the 16-bit Timer1 was used as a timer with no prescaler (or a prescale value of 1:1), then at a clock rate of 8 MHz, this timer will overflow every 32.768 ms. If a 1:8 prescaler was used, then the overflow interval is 262.144 ms.

The PIC-C compiler has functions for setting and reading the timers available on the chip. **To set-up Timer0** for example, *the setup_timer_0()* function is called. The function gives the user choice of clock source (internal or external) and the prescale factor to use. As an example, to use the internal clock on the MCU with a prescale factor of 8, the function is called as

Setup_timer_0 (RTCC_INTERNAL, RTCC_DIV_8)

where the particular parameters used are dependent on the selected MCU (done by adding the header file for the MCU used to the C-language file). Once the timer is set up, **the timer is accessed** by using the provided *get_timer0()* function, which returns the count value of the counter associated with this timer. The compiler also provides the *set_timer0()* function, which sets the count value of the counter to a particular value.

An **interrupt** is a mechanism in which a predefined event (such as the elapse of a hardware timer or the arrival of a character in a serial line) causes a program to stop execution after the current instruction, saves the state of the program, and then executes a predefined routine called an Interrupt Service Routine (ISR). After the ISR completes execution, the program resumes its operation at the next instruction. On the PIC16F690, several sources can cause interrupts. These include:

Timer0 or Timer1 Overflow

PORTA or PORTB change

EUSART Receive and Transmit

External Interrupt on pin RA2

Interrupts are used in order to not waste the computing resources by checking if an event occurs. The process of continuous checking is called **polling**. An example of a polling operation is the process of continuously reading a counter to check if the counter has overflowed. The CCS compiler provides several functions for interrupt processing. These include *disable_interrupts()*, which disables a specified interrupt, and *enable_interrupts()*, which enables a specified interrupt. The *#int_xxx* directive (where *xxx* is the specified interrupt name) is placed before the code listing for the interrupt service routine to inform the compiler to use the following function with interrupts associated with the specified interrupt name. In this laboratory, we will explore three different interrupts. These include a timer overflow interrupt, a capture interrupt, and an RS-232 receive data available interrupt.

Lab 10 Exercises:

Name(s): _____

10.1 The code shown in Figure E10.1 uses **Timer1 on the PIC16F690**. The code has two functions: *SetupTimer()* and *ReadTimerNow()*. The *SetupTimer()* function sets Timer1 as the timer. Timer1 is set to use the internal clock as the timing source with no prescaler. Using an 8 MHz clock, the timer resolution is 0.5 μs. The *ReadTimeNow()* function returns the time in integer units (which are multiples of the timer resolution) since the *SetupTimer()* function was called). The *ReadTimeNow()* function has code to handle the possibility that the Timer1 counter (*LastCount* variable) has overflowed since the last read. Note that, because the *Time* variable is a 32-bit unsigned integer, the *ReadTimeNow()* function can read time correctly up to 2147 seconds ($2^{32} \times 0.5$ μs) unless we use a larger prescale factor. **Use the *ReadTimeNow()* and *SetupTimer()* functions to implement a simple PIC-C program that turns ON LED1 (see Figure 7.5) for 5 seconds and then OFF for 10 seconds in a repetitive fashion. Do not use any of the compiler provided time delay functions to do timing for this exercise.** Attach a printout of the code you developed for this lab.

Figure E10.1

```
/////////////////////////////////////////////////////////////////////
///   Code for setting up a timer in software
///
///     Compiler: PCWH from CCS, Inc. (Version 4.103)
/////////////////////////////////////////////////////////////////////
void SetupTimer(void)
{
    setup_timer_1(T1_INTERNAL);          //At 8 MHz internal clock and 1:1 prescale, the timer
                                         // has a resolution of 0.5 μsec

    TimerRes = 0.0000005;
    LastCount = 0;
    Time = 0;
    set_timer1(0);
}
int32 ReadTimeNow(void)                  //Returns time in units that are multiple of the timer resolution
{
int16 ReadTime;
ReadTime = get_timer1();
if (ReadTime > LastCount)
    Time = (ReadTime - LastCount) + Time;
else
    Time = Time + ((65536 - LastCount) + ReadTime);

LastCount = ReadTime;
return(Time);
}
```

Hint: You need to add a *main()* function that calls the *ReadTimeNow()* and *SetupTimer()* functions.

Troubleshooting Tip: Note that the variables *Time* (int32), *LastCount* (int16), and *TimerRes* (float32) have to be declared as global variables (declared outside of the *main()* function) for the provided two functions to work.

10.2 The PIC-C code listing in Figure E10.2 is for a program that demonstrates **Timer0 overflow interrupt**. The code flashes an LED ON and OFF every two seconds. The timing interval is kept by a variable (*int_count*) that is decremented whenever the interrupt occurs. Note how the *int_count* variable is initialized to the desired number of interrupts at the start of the *main()* function and whenever *int_count* reaches zero. The frequency of the interrupts is controlled by setting the timing parameters for Timer0. In the code shown for an 8 MHz clock and a prescale value of 256, the Timer0 clock frequency is 7812.5 Hz. Since the interrupt occurs when the 8-bit Timer0 counter overflows (i.e., when the count goes from 255 to 0), then the interrupts are generated at the rate of 30.52 interrupts per second or (7812.5/256).

Figure E10.2

```
/////////////////////////////////////////////////////////////////
//                      Timer0int.c
//
//  A program to illustrate timer0 overflow interrupt
//  Compiler: PCWH from CCS, Inc. (Version 4.103)
/////////////////////////////////////////////////////////////////
#include <16F690.h>
#fuses INTRC_IO,NOWDT,NOMCLR
#use delay(clock=8M)
#define INT_PER_2SECONDS 61           // ((8000000*2)/(4*256*256))

int8 seconds;                         // Seconds counter
int8 int_count;                       // Number of interrupts left before 2 seconds has elapsed
int8 LED_state = 0;                   // Flag to keep track of the LED state

#int_timer0                           // This ISR function is called every time
void clock_isr()                      // timer0 overflows (255->0).
                                      // For this program this happens 30.5 times/sec (or 61 times/2 sec)
  {
     if(--int_count==0)               // Check if interrupt counter is zero
     {
     seconds = seconds + 2;           // Increment seconds counter
     if (LED_state == 0)              // Turn on LED if it was off
     {
     output_high(PIN_C0);
     LED_state = 1;
     }
     else
     {
     output_low(PIN_C0);              //Turn off LED if it was on
     LED_state = 0;
     }
     int_count= INT_PER_2SECONDS;     //Reload number of interrupts per 2 second
  }

}
void main()
{
  int_count=INT_PER_2SECONDS;
  set_timer0(0);                      // Initialize timer0 to zero
  setup_timer_0(RTCC_INTERNAL | RTCC_DIV_256 );   // Set timer0 to use internal clock with a prescale of 256
  enable_interrupts(INT_TIMER0);      // Enable timer0 interrupt
  enable_interrupts(GLOBAL);          // Activate timer0 interrupt

  while (2 > 1)                       // Start an infinite loop
  {
  ;                                   // Do nothing
  }
}
```

Modify the provided code as follows.

a. Determine the prescale value to use so you can get an interrupt frequency close to 1000 interrupts per second. Modify the code to incorporate this new value and write the value below.

Prescale value _____

b. Instead of flashing an LED, make the interrupt service routine (ISR) turn ON a digital output pin at the beginning of the ISR and turn OFF that pin just before exiting the ISR. Connect the digital pin to an oscilloscope and measure the period of the displayed signal.

Use the measured period to compute the interrupt frequency. Write down the period below.

Measured period _____

Does the measured interrupt frequency agree with the specified interrupt frequency?

_____

Print a copy of your modified code to give to your instructor.

> **Troubleshooting Tip:** Increase the oscilloscope intensity setting so you can observe the small duration pulse.

_____

10.3 The code shown in Figure E10.3 demonstrates the **capture interrupt** of the capture compare PWM (CCP) module on the PIC16F690. The PIC16F690 has one CCP module called CCP1. The capture feature allows the current reading of Timer1 to be copied to the CPP_1 register when a designated event occurs on the CCP1 pin. In the code listing, the capture mode is set to record the Timer1 reading in register CCP_1 at each rising edge of the signal connected to the CCP1 pin (pin #5). In the capture ISR shown, the CCP1 register value (address defined in header file) is copied to a variable called *SignalEdgeTime*. The main routine displays this variable to the user using a serial RS-232 connection.

a. Compile and test the operation of this code. Use the function generator to supply a square-wave signal at a frequency of 0.1 kHz with a maximum voltage value of 5 VDC and a minimum voltage value of 0 V. Connect the square-wave signal to the CCP1 pin. Run the given code and explain below what you observed.

_____

_____

_____

b. Modify the given code so you can **measure the period of the square-wave signal** connected to the CPP1 pin. Display the measured period (alternatively, the frequency) of the signal to the user using the RS-232 interface. Vary the signal frequency on the function generator, and ensure that your code displays the correct signal period or frequency. Print and attach a copy of the revised code.

```
///////////////////////////////////////////////////////////////////
//          CaptureInt.c
//
// A program to illustrate capture interrupt
// Compiler: PCWH from CCS, Inc. (Version 4.103)
///////////////////////////////////////////////////////////////////
#include <16F690.h>
#fuses INTRC_IO, NOMCLR, NOWDT
#use delay (internal=8M)
#use rs232(baud=38400, BITS = 8, PARITY = N, xmit=PIN_B7, rcv=PIN_B5)

int16 SignalEdgeTime;                 // Value in CPP1 register

#int_ccp1                             // ISR that operates on signal connected to CPP1 pin
void isr()
{
   SignalEdgeTime = CCP_1;            // CCP_1 is the time the pulse went high
}

void main()
{
   setup_ccp1(CCP_CAPTURE_RE);        // Configure CCP1 to capture on rising edge
   setup_timer_1(T1_INTERNAL);        // Set timer1 to use internal clock at 8 MHz

   enable_interrupts(INT_CCP1);       // Enable interrupts for Capture mode on CCP
   enable_interrupts(GLOBAL);         // Activate Interrupts

   while(TRUE)
   {
      delay_ms(5);                    // Set a time delay of 5 ms
      printf("\r\n%lu us ", SignalEdgeTime/2);  // Display signal rising edge time in microsecond. The divide by 2
   }                                  // factor is because the timer resolution is 1/2 microsecond
}
```

Hint: Compute the difference between successive rising edge timer readouts to obtain the signal period.

10.4* The code shown in Figure E10.4a demonstrates **RS-232 receive data available interrupt** on the PIC16F90. Arrival of a character on the RS-232 line causes the ISR to run and to transfer the received character from the input buffer to a storage buffer. The main program then can read available characters in the storage buffer when it has time. In the provided code, the program stores a maximum of *BufferSize* characters in the storage buffer. If more characters were received, the program ignores them. Notice the use of two indices: one to specify where characters are stored and the other to specify where characters are retrieved.

a. Compile and test the operation of this code. Change the *BufferSize* to 51, and notice the increase in the number of characters that can be received. Change the delay interval in the main routine, and comment on its effect on program operation.

Figure E10.4a

```
/////////////////////////////////////////////////////////////////////
//              SerialInt.c
//
// This program illustrates RS232 char available interrupt
// Compiler: PCWH from CCS, Inc. (Version 4.103)
/////////////////////////////////////////////////////////////////////
#include <16F690.h>                      // Include file for the particular chip used
#use delay(internal=8M)                  // Use Internal 8 MHz- clock
#fuses INTRC_IO, NOMCLR                  // Set clock mode to internal oscillator with
                                         // no clkout. Master clear pin is used for I/O
#use rs232(baud=9600, BITS = 8, PARITY = N, xmit=PIN_B7, rcv=PIN_B5)
#define BufferSize 11                    // Storage size
char buffer[BufferSize];                 // Array for ring buffer. Zero element is not used
                                         // Note char type is the same as unsigned int8

int8 NumofChar = 0;                      // Counter for number of characters placed in storage buffer
int8 StorageIndex = 0;                   // Index for array element at which char is stored
int8 RetrievalIndex = 0;                 // Index for array element at which char is retrieved
char d;                                  // Variable of type char

#int_rda                                 // Character available ISR
void serial_isr()
{
  d=getc();                              // Retrieve character from input buffer
  if (StorageIndex <= (BufferSize-1))    // In C, the initial array element is the zero element
  {
    buffer[StorageIndex] = d;            // Store character in storage buffer
    StorageIndex++;                      // Increment storage index
    NumofChar++;                         // Increment number of characters placed in storage buffer
  }
}

void main()                             // Main program
{
  char c1;                               // Local variable
  enable_interrupts(int_rda);            // Set RS232 data available interrupt
  enable_interrupts(GLOBAL);             // Activate interrupts

  while(TRUE)                            // Start an infinite loop
  {
    delay_ms(1000);                      // Set a time delay of 1000 ms
    if (NumofChar > 0 )                  // Check if there is still unread characters in storage buffer
    {
      c1 = buffer[RetrievalIndex];       // Retrieve char from storage buffer
      RetrievalIndex++;                  // Increment retrieval index
      printf("%c \n\r", c1);             // Send char with new line and CR characters
      NumofChar--;                       // Decrement number of characters available to read
    }
  }
}
```

b. Modify the provided code in Figure E10.4a to allow for **unlimited storage and retrieval of received characters**. An efficient way to achieve this is through the use of **a ring-type storage buffer**. A ring buffer is implemented as an array of characters with two pointers called *BufferHead* and *BufferTail* that control the storage and retrieval of items in the buffer. Items are stored at the current location of the *BufferTail*, while items are retrieved at the current location of the *BufferHead*. After a data item is stored in the buffer, the *BufferTail* pointer is incremented. Similarly, after a data item is retrieved from the buffer, the *BufferHead* pointer is incremented. At startup or when the buffer is empty, the two pointers point to the same location.

A ring buffer allows a wraparound, thus the name ring buffer. To organize the storage, four functions should be used. These are listed here.

int8 IsBufferEmpty(void);	// Function to check if buffer is empty
int8 IsBufferFull(void);	// function to check if buffer is full
void StoreItemInBuffer(char);	// Function to store an item in the buffer
char GetItemFromBuffer(void);	// Function to get an item from the buffer

Figure E10.4b is a code listing for two of these functions. Write the code for the remaining two functions, and incorporate these functions into the provided code so you can use a ring buffer for storage and retrieval of characters. Note that, when using a ring buffer, the effective storage size is *BufferSize-1*.

Print and attach a copy of the code you developed.

Troubleshooting Tip: You should check if the buffer is not empty before retrieving items from the buffer. Also the BufferTail and BufferHead pointers should be initialized to 1 at the start of the program.

```
/////////////////////////////////////////////////////////////////////
//   Code listing for the functions IsBufferFull and GetItemFromBuffer
//
// Compiler: PCWH from CCS, Inc. (Version 4.103)
/////////////////////////////////////////////////////////////////////

// Note in this code, we assume that the storage buffer starts at element #1 of the array and
// element #0 is not used for storage

int8 IsBufferFull(void)                  // Returns 1 if buffer is full and zero otherwise
  {
  int8 answer;
  if (BufferTail == (BufferSize-1))      // Is the BufferTail at the end of the array?
    {
    if (BufferHead == 1)                 // Buffer is full if the BufferHead points to array element #1
      answer = 1;
    else
      answer = 0;                        // Buffer is not full since the BufferHead is not at the top of the array
    }
  else
    {                                    // Code for case where the BufferTail is not at the end of the array
    if ((BufferTail + 1) == BufferHead)  // Buffer is full if the BufferHead is at the location after the
                                         // BufferTail
      answer = 1;
    else
      answer = 0;
    }
    return(answer);
  }

char GetItemFromBuffer(void)             // Retrieve character from ring buffer
{
  char item;
  Item = Buffer[BufferHead];
  if (BufferHead == (BufferSize-1))      // Adjust the BufferHead pointer after the char was retrieved from
                                         // the buffer
    BufferHead = 1;                      // Wrap around to item 1 again if the end if the array was reached
  else
    BufferHead = BufferHead + 1;         // Increment the BufferHead pointer
  return (item);
}
```

QUESTIONS

10.1 What is the advantage of using timers instead of time-delay functions?

10.2 What changes do you need to make to the *SetupTimer()* function in Lab 10.1 to obtain a timer with a coarser resolution? What is the largest resolution that can be obtained?

10.3 Does the PIC16F690 support interrupt priorities?

10.4 List one or more applications that the capture feature on the PIC16F690 can be used for.

Timing Functions in a PC and State Transition Diagrams

After you complete this lab, you should be able to:

- Know how to use the different timers available in Visual Basic Express (VBE)
- Organize the operation and control of physical systems into tasks and states
- Implement state-transition diagrams on a PC and/or MCU
- Develop a graphical user interface (GUI) using VBE

For this lab you need:

- RS-232 interface circuit between a PIC MCU and a PC or a PIC development board with a built-in RS-232 interface
- A serial cable or a USB-to-serial cable (Sewell Direct SW-1301)

LAB BACKGROUND/INFORMATION:

Lab 10 covers timers in a PIC MCU. The **first part of this laboratory covers timers available on a PC platform** and uses the timing functions available in Visual Basic Express (VBE) as an illustration. The **second part of this laboratory deals with control software development based on the task/state structure**. This laboratory also addresses the development of graphical user interfaces on a PC.

VBE has several functions that access the counters that are kept by the operating system. In this laboratory, we will make use of two timing functions. These are

- Timer property
- Timer component

The **Timer property** returns the number of seconds since midnight, while the **Timer component** allows a Windows-based application to respond to events that are spaced regularly. We also will make use of the **Performance Counter** in this laboratory to implement timers with sub-microsecond resolution.

Software development is a very important piece in the development of a mechatronic system. Without software, an MCU or PC cannot function. A software-based control system offers flexibility over a hardware-based one, since the controller structure and control logic can be changed by simply changing the code in the program. Due to the different types of control activities that need to be performed (such as feedback control or discrete event control), it is advantageous to develop a uniform control software structure that can handle a variety of control applications. This laboratory applies a structure that is based on the **task/state** software structure. In this structure, the operation of a physical system is broken into one or more tasks. A **task**

is a collection of software activities to be performed. Within a task, the code is organized into separate states. Each **state** signifies a distinct condition of operation of the system or machine. A system or machine can be in only one state at a time. The transitions between states are shown using **state-transition diagrams**.

As an illustration of task/state organization, let us create a state-transition diagram for a program that needs to generate the following periodic signal (see Figure 11.1). The program should start sending this signal pattern in response to a *Start* command and should shut off the output in response to a *Stop* command.

Figure 11.1

Periodic on/off signal

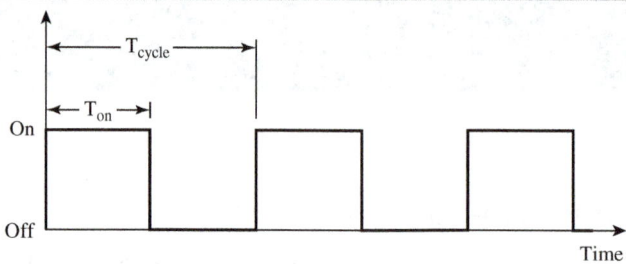

The **state-transition diagram for this task** is shown in Figure 11.2. We have broken the operation of this simple program into three states shown as rectangular blocks in the figure. The conditions that cause transitions between these states are shown in italics along the arrows that connect the blocks. The program starts in the *Initial* state waiting for the *Start* command to be issued by the user. Once the *Start* command is received, the program switches states to the *Output_On* state, where the output is turned on. The program remains in this state while the elapsed time (T), from the beginning of each cycle, is less than or equal to T_{on}, which is the on-time interval. When the elapsed time exceeds T_{on}, the program switches to the *Output-Off* state, where the signal is switched OFF and remains in this state while the elapsed time is less than or equal to T_{cycle}, which is the cycle time. The program cycles between the *Output_On* and *Output_Off* states based on timing signals, unless the *Stop* command was issued, which causes the program to go back to the *Initial* State and wait for a new *Start* command.

Figure 11.2

State-transition diagram for generating a periodic signal

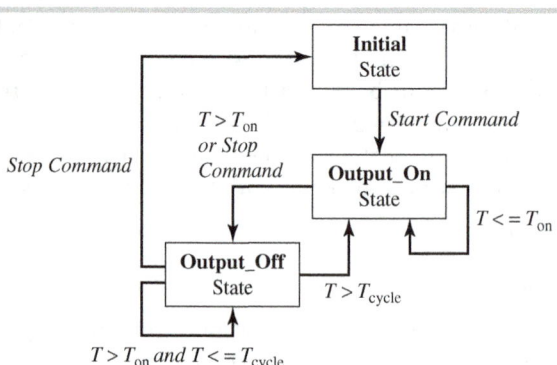

We will consider in this laboratory software implementation of state-transition diagrams using the **cooperative control mode**. In this mode (see Figure 11.3), we set up an infinite Do-loop that runs indefinitely while the *Stop* variable is not equal to 1. Within the infinite loop, we include functions to call up the tasks that we want to run. In this example, we consider three tasks. When a task is called, the code in the active state in that task is executed. Once the code completes execution, the next task on the list is called, and the process repeats. The last three lines inside the loop in Figure 11.3 contains the code that allows background processing. This code is necessary if the code is implemented in a Windows operating system to allow Windows to process any background events that have occurred during the execution of this code.

Figure 11.3

```
While (Stop not equal to 1)
{
        Increment Count
        Call Task1()
        Call Task2()
        Call Task3()
        If (Count mod 1000) = 0 then
                Call function to allows background processing
        Endif
}
```

Pseudocode for implementing the cooperative control mode in software

This laboratory includes exercises on how to use the PC timing functions and on how to write control software using the task/state software structure. It includes PC- and MCU-based exercises. The PC exercises use Visual Basic Express (VBE) as the programming language, while the MCU-based exercises use the PIC-C language. The PC exercises demonstrate the development of **graphical user interfaces (GUIs)** on a PC.

LAB 11 EXERCISES:

Name(s): _____

11.1 This exercise uses the **Timer property in VBE to return time information**. The Timer property returns the number of seconds since midnight, and it is called using the syntax *Microsoft.VisualBasic.Timer*. The GUI for a VBE application that uses the Timer property is shown in Figure E11.1a. This Windows form application uses two command buttons, four textboxes, and two group boxes. Pressing the button *GetTimeSinceMidnight* returns the time (in seconds) since midnight, while pressing the button *GetTimeSinceStart* returns the time (in seconds) since the application was started. Create this Windows application in VBE and enter the code listing shown in Figure E11.1b for this application. Change the name property for the *GetTimeSinceMidnight* button to *cmdGetTimeMidnight*, and the name property for the *GetTimeSinceStart* button to *cmdGetTimeStart*. The *Form1_Load* function in Figure E11.1b is a function that automatically executes when the application is loaded. This routine, which executes only once, is used to store the value of the *Timer property* at the start of the application.

Figure E11.1a

GUI for Exercise 11.1

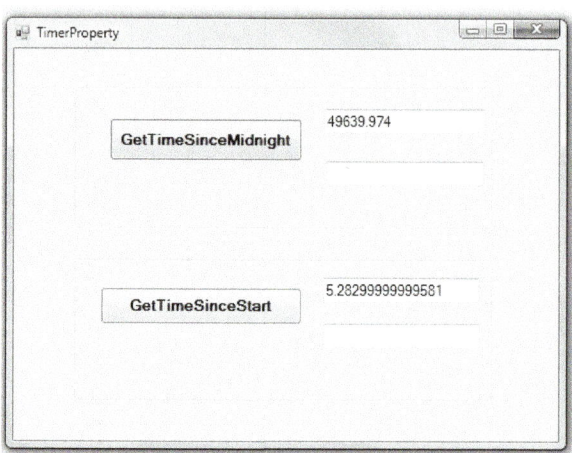

Figure E11.1b

VBE code for Exercise 11.1

```
Public Class Form1

    Dim StartTime As Double

    Private Sub cmdGetTimeMidnight_Click(ByVal sender As System.Object, ByVal e As System.EventArgs)
    Handles cmdGetTimeMidnight.Click
        TextBox1.Text = Microsoft.VisualBasic.Timer
    End Sub

    Private Sub cmdGetTimeStart_Click(ByVal sender As System.Object, ByVal e As System.EventArgs) Handles
    cmdGetTimeStart.Click
        TextBox3.Text = Microsoft.VisualBasic.Timer - StartTime
    End Sub

    Private Sub Form1_Load(ByVal sender As Object, ByVal e As System.EventArgs) Handles Me.Load
        StartTime = Microsoft.VisualBasic.Timer
    End Sub

End Class
```

Run the program by pressing *Debug\Start Debugging*, and make sure that the application operates as described. Revise the code to do the following.

a. Display the time in hours in the blank textbox associated with each command button (*TextBox2* in the upper group box and *TextBox4* in the lower group box).

b. Add two new textboxes (one to each group area) and the corresponding code to display the number of times the command button associated with that textbox was pressed.

Print a copy of your code plus a screen shot of the program while in operation.

> **Troubleshooting Tip:** The variable used to keep track of the number of times the command button was pressed should be declared as a global variable or as a local variable using the *Static* qualifier in the declaration statement.

11.2 **Implement a program that operates as a stopwatch timer using VBE.** Use the *Timer component* in VBE for timekeeping. The GUI design for the stopwatch timer application is shown in Figure E11.2. The GUI has four command buttons, two labels, two textboxes, and one group box. At startup, both the *Start* and the *Stop* commands should be disabled, and only the *Exit* and *Run Timer* commands should be enabled. The function of each command button is explained here.

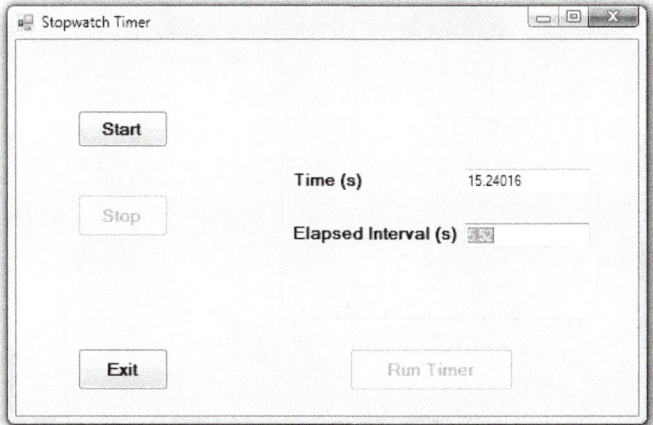

Figure E11.2

GUI for Exercise 11.2

Run Timer: This command should set up the interval for and enable the Timer component. Enabling the Timer component should cause the **timer tick routine** associated with the Timer component to start running to display the time (in seconds) in the *Time* textbox since the *Run Timer* command was issued. In addition, the command should enable the *Start* button and disable the *Run Timer* command button.

Start: This command should cause the stopwatch to start operating (using the timer tick routine) to display the elapsed time in the *Elapsed Interval* textbox since this command was pressed. This command also should cause the *Stop* button to be enabled and the *Start* button to be disabled.

Stop: This command should cause the stopwatch timer to stop. This means that the update of the elapsed time information will be stopped. This command also should cause the *Stop* button to be disabled and the *Start* button to be enabled.

Exit: This command should cause the program to end. This will cause the *Stopwatch Timer* form to close.

Run your code and ensure that it operates as specified. Print a copy of your code plus a screen shot of the program while in operation.

> **Troubleshooting Tip:** You need to use only one timer tick routine to solve this exercise. For accurate time keeping, the Timer component interval should not be smaller than 50 ms.

11.3 Figure E11.3a shows the GUI design while Figure E11.3b shows a screen shot of the application in operation for a **VBE program that uses the task/state control software structure**. The code listing for this application is shown in Figure E11.3c. This application implements **a single software task that has two states**. In state 1, the code waits for 5 seconds to elapse before transitioning to state 2, which waits for 10 more seconds before going back to state 1. The process repeats indefinitely. The task is called repeatedly inside an infinite loop, which is scanned continuously once the user hits the *Start* button. Note the use of the variables *State* and *NextState* to implement state transitions. Implement this application in VBE, and make sure that the program operates as described.

Figure E11.3a

GUI for Exercise 11.3

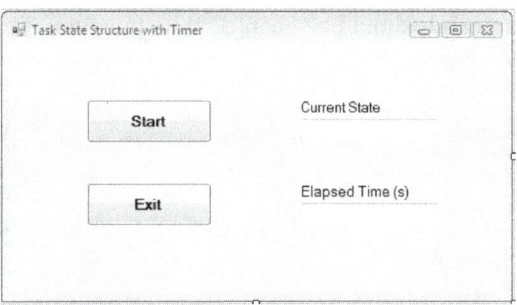

Figure E11.3b

Screen shot of program

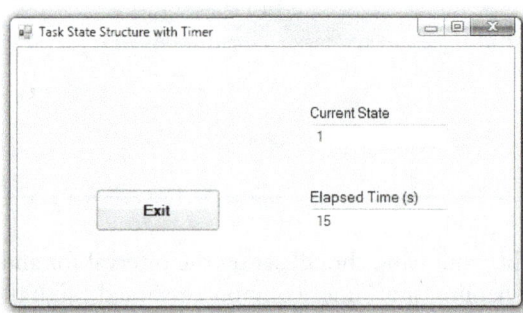

This application uses the *Timer property* in VBE for timekeeping, which has a resolution of about 15 ms. **Revise this code to use the *Performance Counter* for timekeeping**, which has a sub-microsecond resolution. Specifically, do the following.

a. Add the class file *PerformanceTimer.vb* (available from the text website) to your project (use *Add/Existing Item* in Solution Explorer).

b. Declare a global variable called *tmr* of type *PerformanceTimer*. Add code to initialize the *PerformanceTimer* in your code (see *PerformanceTimer.vb* code).

c. Replace the contents of the *GetTimeNow()* function with code that uses the *PerformanceTimer* instead of the *Timer property*.

Print a copy of the revised code.

```vbnet
Public Class Form1
'***************************************************************
' Task State Struture with Timer
' A program to illustrate implementation of task state software
' struture using cooperative scheduling
'***************************************************************
                                  ' Global variables declaration
Dim State, NextState As Integer   ' State and NextState variables
Dim LastTime As Double            ' Variable to store last time information
Dim ElapsedTime As Double         ' Time since application was started
Dim k As Long                     ' Index for number of scans
Private Sub Form1_Load(ByVal sender As System.Object, ByVal e As System.EventArgs) Handles MyBase.Load
    TextBox1.Text = " "           ' Initialize TextBox1 display information
    TextBox2.Text = "0"           ' Initialize TextBox2 display information
    NextState = 1                 ' Set initial state to 1 at startup
End Sub

Public Function GetTimeNow() As Double
    'This function uses the Timer Property for time keeping which has a resolution of about 15 ms
    GetTimeNow = Microsoft.VisualBasic.Timer()
End Function

Public Sub ExampleTask()
    State = NextState
    Select Case State

        Case 1 '  //State 1
' This state simply waits for 5 seconds before it jumps to state 2
            If ((GetTimeNow() - LastTime) >= 5) Then
                ElapsedTime += 5
                LastTime = GetTimeNow()
                NextState = 2
            End If

        Case 2 ' //state 2
' This state simply waits for 10 seconds before it goes back to state 1
            If ((GetTimeNow() - LastTime) >= 10) Then
                ElapsedTime += 10
                LastTime = GetTimeNow()
                NextState = 1
            End If
    End Select
End Sub

Public Sub UpdateCurrentState()
    TextBox1.Text = Str$(State%)
    TextBox2.Text = Str$(ElapsedTime#)
End Sub
Public Sub UpdateCurrentState()
    TextBox1.Text = Str$(State%)
    TextBox2.Text = Str$(ElapsedTime#)
End Sub

Private Sub cmdStart_Click(ByVal sender As System.Object, ByVal e As System.EventArgs) Handles cmdStart.Click
    cmdStart.Visible = False            ' Hide the Start Command
    LastTime = GetTimeNow()             ' Record the starting time value
    Do While (2 > 1)                    ' This is an infinite loop
        k = k + 1
        If k Mod 3000 = 0 Then ' The mod function is used so that the following
            ' two statements do not get executed all the time
            System.Windows.Forms.Application.DoEvents() ' Allows background processing
            Call UpdateCurrentState()
        End If
        Call ExampleTask() ' This task code function is called once in every scan through the loop
    Loop
End Sub

Private Sub cmdExit_Click(ByVal sender As System.Object, ByVal e As System.EventArgs) Handles cmdExit.Click
    End                                 'End the program and close the form
End Sub

End Class
```

> **Hint:** For flexibility, add a flag to your code that allows you to select the timer source (*Timer property* or *Performance Counter*). As long as you use the *GetTimeNow()* function to get time information, the timing source is hidden from the user, and only the timer resolution changes.

11.4 **Implement a program using VBE** that uses the task/state structure to **simulate the operation of a linear positioning stage**. The stage operation should be controlled by five commands. The function of each is explained here.

Start: This command should cause your program to start monitoring for user commands but should not cause the stage to move. The command should disappear (or be inactive) after you have pressed it once. All other commands (with the exception of the *Exit* command) should have no action until you have pressed the *Start* command.

Right: This command should cause the stage to move to the right. The stage should automatically stop when it touches the right limit switch. If the stage is at the right limit switch location, then this command should do nothing.

Left: This command should cause the stage to move to the left. The stage should automatically stop when it touches the left limit switch. If the stage is at the left limit switch location, then this command should do nothing.

Stop: This command should cause the stage to stop moving if it was moving, but the program should still be running.

Exit: This command should first stop the stage if it was moving and then exit the control program.

This program should allow the user to vary the travel speed of the stage. In addition, when the stage hits either of the limit switches, means must be provided to inform the user of this situation. A suggested VBE GUI for this problem is shown in Figure E11.4a. The GUI uses five command buttons, one for each of the previous commands. The stage position is displayed in a textbox. The GUI uses radio buttons to allow the user to select one of three different speed settings for the stage motion. In addition, the relevant label will be made visible when the stage hits either limit switch.

Figure E11.4a

GUI for Exercise 11.4

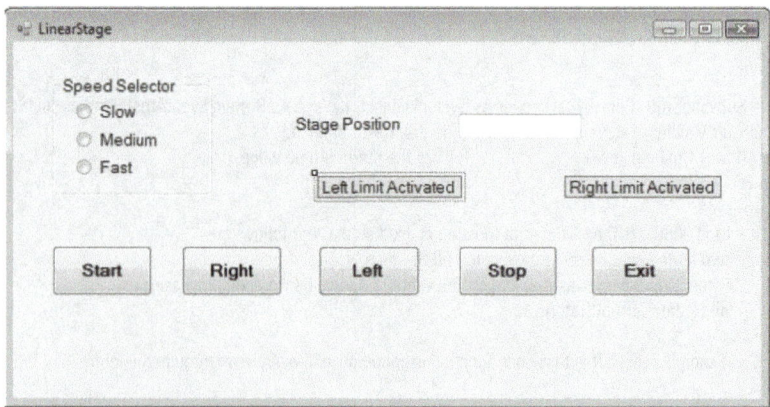

To give flexibility for converting this simulation program into a real implementation at a later point, the control software should be split into **two separate tasks** that are run in cooperative fashion. One task (call it *Sequence Task*) should be made to process the referenced commands, while another task (call it *Motion Task*) takes care of simulating the motion of the stage. The two tasks are called repeatedly inside an infinite loop that is called when the user hits the *Start* command. Refer to the code example given in Exercise 11.3 to help you in setting up the code structure. As a first step in solving this lab, draw an appropriate state-transition diagram for the operation of each of the two control tasks. Note that all of the hardware/simulation details should be isolated from the *SequenceTask*. For example, to move the positioning table to the right, the *SequenceTask* can use functions to set the *MotionStatus* variable to *ON* (or 1) and the *MotionDirection* variable to *Right* (or 1). The *MotionTask* accesses these variables to simulate the motion of the stage.

For simulation of the motion of the stage, the *MotionTask* increments the stage position if the stage is moving to the right and decrements the stage position if the stage is moving to the left. The stage position variable starts at 0. The incrementing/decrementing rate should be controlled by a timer which makes use of the user-selected speed setting. To simulate the limit switch, the stage is considered to be at the limit switch if the position variable reaches a user-specified limit value (such as $+/-1000$). The *SequenceTask* can determine if the stage has reached the limit by interpreting the stage position information. A screen shot of the suggested GUI when the stage has reached the right limit is shown in Figure E11.4b.

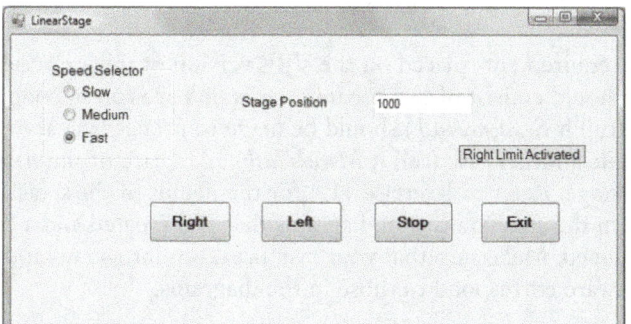

Figure E11.4b

Screen shot of program in operation

For this lab exercise, hand in the state-transition diagrams that you created, a copy of the code you developed, and multiple screen shots of the program while in operation. Make sure that your task break-up into states and their transition in the software correspond to those in the diagrams.

> **Hint:** Use a variable called *Command* that is set to a particular value when the user clicks on any of the command buttons. The *SequenceTask* uses the value of this variable in making decisions on transitions between states.

11.5 **Implement a program in a PIC development board** that uses the task/state structure **to simulate the operation of a linear positioning stage** similar to that described in Exercise 11.4. Since a PIC MCU does not have a built-in GUI, use the elements shown in Figure E11.5 for input and display. The *Right*, *Left*, and *Stop* commands are entered using push-button switches connected to available digital-input pins on the MCU. There is no need for a *Start* and an *Exit* command when using the PIC MCU. A rotary pot is used to vary the stage speed, while the activation of the limit switches is indicated by turning on LEDs. The position of the stage is displayed using a PC terminal program connected to the PIC MCU through the serial port.

Figure E11.5

Interface elements for Exercise 11.5

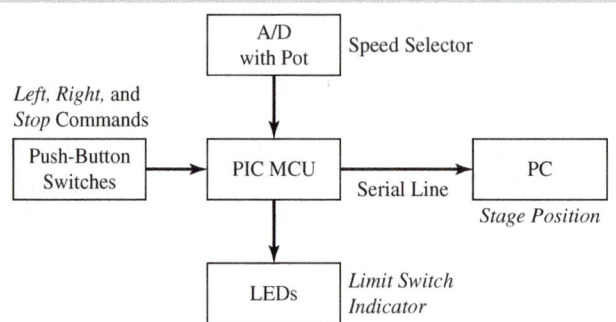

Similar to the requirements placed on the VBE version of this exercise, your software solution should consist of two separate tasks that are run in cooperative fashion. One task (call it *SequenceTask*) should be made to process the above-referenced commands, while another task (call it *MotionTask*) takes care of simulating the motion of the stage. Refer to Exercise 11.4 for the details of these tasks. For this exercise, hand in the state-transition diagrams that you created and a copy of the code you developed. Make sure that your task break-up into states and their transition in the software correspond to those in the diagrams.

> **Troubleshooting Tip:** Make sure to declare the correct variable type (size and signed or unsigned) for the variables used in this program. Note that, in the PIC-C compiler, all integer type variables are by default unsigned.

11.6* **Implement a program in VBE that simulates the operation of an oven.** The oven should use a simple ON/OFF controller for the heating element. A suggested GUI for this problem is shown in Figure E11.6. The GUI uses four command buttons, two textboxes, five labels, one group box, and one numeric up/down control.

Figure E11.6

GUI for Exercise 11.6

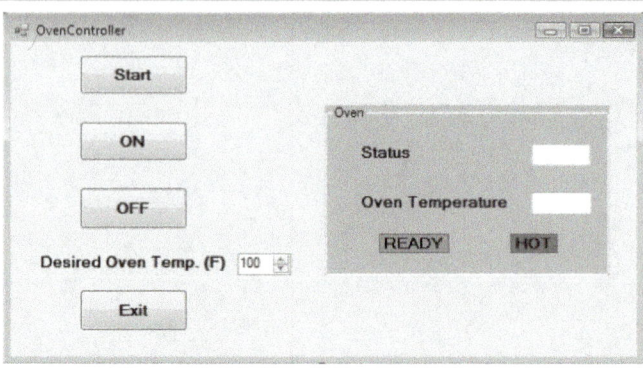

The oven should operate according to the following rules.

a. The program does not respond to commands until the user has hit the *Start* command. The *Start* command should be made invisible (or disabled) after it has been pressed once.

b. The heating element is turned ON if the command *ON* is pressed *and* the desired oven temperature is above the current oven temperature.

c. The heating element is turned OFF when the command *OFF* is pressed.

d. The oven heating element should be turned off automatically when the oven temperature exceeds the desired oven temperature.

e. The oven READY label should be made visible whenever the oven temperature is equal to or exceeds the desired oven temperature.

f. The HOT label should be made visible whenever the oven temperature is above 120° F.

g. When the heating element is ON, the oven temperature should increase at the rate of 2°/s.

h. When the heating element is OFF, the oven temperature should decrease at the rate of 1°/5s.

i. At startup, only the *Start* and *Exit* commands are visible, and the READY and HOT labels should be invisible.

As a first step in solving this exercise, create a state-transition diagram to handle the operation of this oven according to the rules. Then use the task/state structure for implementing this oven simulator (refer to the code structure given in Exercise 11.3). For this exercise, you could use one task to handle the oven's states of operation and user commands and the *Timer component* for simulating the heating/cooling action. The *Timer component* was used in Exercise 11.2, so refer to that for further details. The *Timer component* provides a simple way of automatically executing a routine at a desired rate. Set the *Timer component* interval to 500 ms. In the timer tick routine associated with that timer, increment the temperature variable by 1° in every call of the timer tick routine if the heating element is ON, and decrement the temperature variable by 1° in every 10 calls of the timer tick routine if the heating element is OFF.

For this exercise, hand in the state-transition diagram that you created, a copy of the code you developed, and multiple screen shots of the program while in operation.

Hint: The oven start temperature should be the room temperature or 68°F (20°C). Also, when the oven is cooling after it was turned off, the lowest temperature it reaches should be the room temperature.

11.7* **Implement a program in a PIC MCU to simulate the operation of an oven** with commands and rules similar to those described in Exercise 11.6. Since a PIC MCU does not have a built-in GUI, use the elements shown in Figure E11.7 for input and display. The *ON* and *OFF* commands are entered using push-button switches connected to available digital-input pins on the MCU. There is no need for a *Start* button. A rotary pot is used to vary the desired oven temperature, while the activation of the HOT and READY labels is indicated by turning on LEDs. The desired and actual oven temperatures are displayed using a PC terminal program connected to the PIC MCU through the serial port.

Figure E11.7

Interface elements for Exercise 11.7

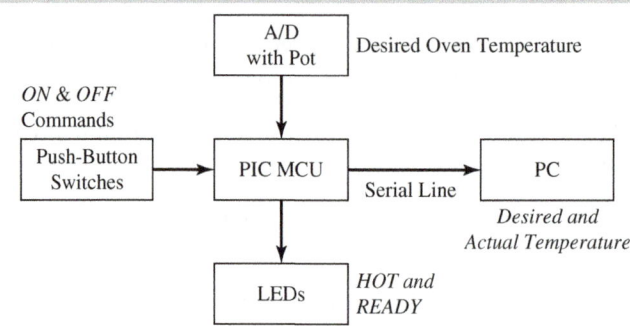

For this exercise, **you can use two tasks running in a cooperative fashion** to solve this problem. One task (call it *Control*) will handle the oven states and user commands, while the other task (call it *HeatingCooling*) will be a timed task that simulates the heating/cooling actions. Alternatively, the heating/cooling actions can be implemented using a **timer-overflow Interrupt Service Routine (ISR)** (see Exercise 10.2). In this case, the *Control* task will be scanned repeatedly inside an infinite loop while the *HeatingCooling* task will execute asynchronously at a specified timing rate as part of the ISR. The oven temperature will be increased (decreased) inside the ISR at a specified rate (see Exercise 11.6) depending on whether the heating element is ON (OFF). The heating element is a global variable that is set or reset inside the *Control* task and whose value can be accessed by the ISR. Note that the transmission of the desired and actual oven temperature to the terminal program should be done by the *Control* task in order to not waste the ISR execution time in this activity. For this exercise, hand in the state-transition diagrams that you created along with a copy of a code you developed. Make sure that your task break-up into states and their transition in the software correspond to those in the diagrams.

Hint: The *Control* task can be structured to have just a few states.

Troubleshooting Tip: Make sure to declare the correct variable type (size and signed or unsigned) for the variables used in this program. Note that in the PIC-C compiler, all integer type variables are by default unsigned.

11.8* Develop **a VBE program to test the A/D, D/A, and digital I/O functions on a data-acquisition card**. For testing the A/D and D/A, the program should allow the user to specify which channel to test. In addition, the program should provide three testing modes: single reading from or writing to a particular channel, repeated testing of a particular channel (for reading only), and continuous testing of a combination of an A/D and D/A channels (for example reading a sine wave through the A/D and sending it back through the D/A). In the second testing mode, the program should allow the user to specify the number of times the individual reading needs to be taken and should display the average reading and the range of the readings taken during the test. For testing the digital port, *output* the program should be able to send a value (0 or 1) to any one of the 8-bits on the output digital port without affecting the current values on the other remaining 7 bits. For testing the input digital port, the program should be able to read the input value on any one of 8 bits. (Note: This exercise assumes the availability of a data-acquisition card with a software library for accessing the A/D, the D/A, and the digital I/O ports). For this exercise, hand in a copy of the code you developed along with multiple screen shots of the program while in operation.

> **Troubleshooting Tip:** For the continuous testing mode of the A/D and the D/A use the *DoEvents()* method in VBE (*System.Windows.Forms.Application.DoEvents()*) inside the *Do-loop* that implements this mode to prevent the program from appearing to be locked.

QUESTIONS

11.1 What is the difference between the Timer property and the Timer component in VBE?

11.2 What is needed for several tasks to execute almost simultaneously in a single processor system?

11.3 Name several conditions that cause transitions between states.

Sensors

After you complete this lab, you should be able to:

- Interpret sensor performance specification
- Know the operation of selected sensors
- Interface selected sensors to an MCU

For this lab you need:

- LM35C sensor
- MP101301 Hall-effect sensor (Alternative: Honeywell SS466A)
- TDB05LFPN magnetic buzzer
- Small magnet
- Resistors: 330 Ω and 1 kΩ

LAB BACKGROUND/INFORMATION:

Sensors are vital components of mechatronic systems, since they provide information that allows us to monitor and to control the operation of these systems. Without the availability of sensory information, automated systems cannot operate. A sensor is an element that produces an output in response to changes in a physical quantity (such as temperature, force, or displacement). There are many types of sensors available. This lab will focus on two sensors. These are an IC temperature sensor and a Hall-effect proximity sensor.

The **LM35C temperature sensor** is an integrated circuit (IC) sensor that is based on transistor technology, specifically the fact that the difference in forward voltage of a silicon *pn* junction is directly proportional to temperature. The sensor is available in several packages (see Figure 12.1), including the hermetic TO-46 metal can package and the TO-92 plastic package. The sensor has three leads: one for 4 to 30 VDC input power, another for ground, and the third for the analog voltage output from the sensor. The analog output of the sensor is proportional to temperature in degrees Celsius (°C) with sensitivity of 10.0 mV/°C. The sensor has a temperature measurement range of −40 to 110°C. This sensor is very suitable for use with microcontrollers, since it draws very little current (less than 60 μA), and its output is directly calibrated in degrees Celsius, thus avoiding any conversion operations.

Hall-effect sensors are solid-state sensors that are used for proximity measurement. They are constructed using semiconductor processing techniques. A Hall-effect proximity sensor consists of two pieces: a stationary sensor package and a magnet that

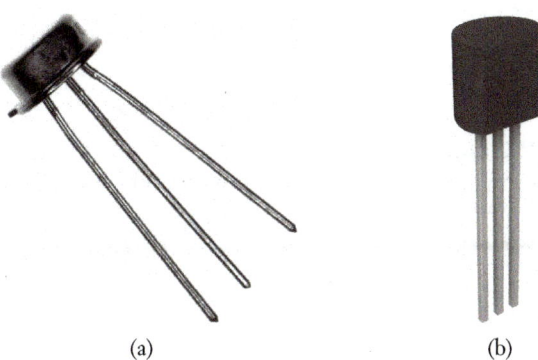

(a)   (b)

Figure 12.1

LM35C sensor (a) TO-46 metal
can package and (b) TO-92
plastic package
(Courtesy of Digi-Key Corporation.)

is attached to the object whose presence needs to be detected, as seen in Figure 12.2. The magnet and the sensor package are separated by an air gap. There are two variations of Hall-effect sensors: unipolar and bipolar. In the **unipolar design**, when a south pole magnet approaches the designated package surface within a specified distance, the sensor turns ON. When the magnet is removed, the sensor turns OFF. In the **bipolar design**, removal of the south pole does not cause the sensor to turn OFF; a north pole needs to approach the sensor to cause the sensor to switch OFF. The Cherry Semiconductor sensor used in this lab is of the unipolar type, while the SS466A sensor is of the bipolar type.

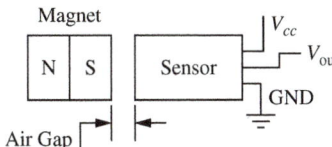

Figure 12.2

Hall-effect proximity sensor

LAB 12 EXERCISES:

Name(s): _____

12.1 Wire up the **LM35C temperature sensor** using 5 VDC as the supply voltage so you can use it to measure the room air temperature. Draw the wiring diagram below.

Use the multimeter to read the output voltage of the sensor. Record the output from the voltmeter. Use the sensor specification to determine the temperature indicated by the sensor and write your findings below.

Voltage reading _____   Indicated temperature_____

Is the indicated temperature reasonable? _____

Connect the output of the sensor to an oscilloscope. Set the oscilloscope input to AC so you can monitor the noise on the sensor output. Measure the noise level and write the value below.

Noise level (mV) _____

From the noise data, can you estimate the repeatability error in using this sensor? Write your estimate below.

Repeatability error (°C): _____

How this value compares with the sensor specification?

_____

Troubleshooting Tip: If the sensor output is very noisy, then you should place a capacitor between the sensor output and the ground lines to filter out the noise. Use a 0.1 μF capacitor or higher.

12.2 Wire up the **Hall-effect proximity sensor** using 5 VDC as the supply voltage. The sensor output should be high when the magnet is not present and should be low when the magnet is present. Determine how close the magnet needs to be to the sensor body for turning the sensor ON or OFF, and write that value below.

Detection distance _____

What is the sensor output voltage when the magnet is present? Write that value below.

Output voltage _____

> **Troubleshooting Tip:** The proximity sensor has an open-collector output circuit and requires a pull-up resistor placed between V_{CC} and the output lines.

Use the proximity sensor to turn ON the buzzer when the magnet is *not* placed close to the sensor to simulate the operation of a single-sensor security system. Draw the wiring circuit for this system below.

> **Troubleshooting Tip:** The buzzer needs 15 to 30 mA current to work properly.

Test the operation of your circuit, and measure the sensor output voltage when the buzzer is ON. Write that value below.

Sensor output voltage _____

Increase V_{CC} from 5 VDC to 10 VDC, and notice what happens to the alarm sound.

> **Troubleshooting Tip:** If you used the SS Hall-effect sensor, then you need to alternate between the north and south poles of the magnet to cause the buzzer to turn ON or OFF.

12.3 Connect the **LM35C sensor output to the A/D input on a PIC MCU**, and develop a program to read the voltage output from the sensor, convert to temperature in engineering units, and then display the temperature reading to the user. You could use a development board with a built-in RS-232 connector or build an RS-232 interface to the PC (see Lab Exercise 9.1), then transmit the temperature reading every second to a terminal program. Attach a copy of the code you developed. Is the variation in the displayed temperature reasonable?

_____

Troubleshooting Tip: If you are using the low pin-count board, then you need to use an A/D pin other than zero.

Touch the sensor with your fingers and comment on the response of the sensor.

_____

_____

12.4* Develop a VBE program that can display the room temperature using the LM35C sensor. Connect the sensor output to one of the A/D channels on the data acquisition card that the PC has. Display the temperature every one second in a textbox. If time permits, use the *Panel* component in VBE to graphically show the time history of the temperature. Hand in a copy of the code you developed as well as a screen shot of the program while operating. (*Note*: This lab exercise assumes the availability of a data-acquisition card with a software library for accessing the A/D).

QUESTIONS

12.1 How can you improve the voltage resolution of a sensor that is read by an A/D converter?

12.2 How can you determine the accuracy of a sensor output?

12.3 What affects the measurement repeatability?

Stepper Motors

LAB BACKGROUND/INFORMATION:

This lab demonstrates the interface and operation of a stepper motor. A **stepper motor** can be classified as a DC motor since it is driven by non-alternating voltages, but its construction and operation are distinct from a DC motor. Stepper motors (as the name suggests) can move in small angular increments (or steps), ranging from 0.9° per step to 90° per step, depending on the construction of the motor and on how it is driven.

There are three types of stepper motors. These are permanent magnet (PM), variable reluctance (VR), and hybrid. **This lab will utilize a PM stepper motor.** In a PM stepper motor, the rotor is a permanent magnet and has no teeth. A PM stepper motor has the capability of exerting a small holding torque, called a **detent torque**, when the stator is not energized due to the use of a magnetized rotor. PM motors are widely used in non-industrial applications (such as computer printers and typewriters). In all configurations of stepper motors, the stator is constructed from pairs of electromagnets commonly referred to as poles.

Stepper motors are classified by the number of phases they have and the polarity (unipolar or bipolar) of the excitation voltage. A **phase** refers to a coil winding; thus, a two-phase motor has two separately activated coil windings, and the coil windings are placed perpendicular to the rotor. The motion of the motor depends on how the stator coils or phases are actuated. There are **four possible ways of actuating the phases**. These are wave drive, full stepping, half-stepping, and micro-stepping.

LAB 13 EXERCISES:

Name(s): _____

13.1 Measure the resistance between the different leads in the **six-lead**, **small four-phase stepper motor** to identify the wire colors of the A, B, C (or $\overline{A}$), and D (or $\overline{B}$) phases of the motor. List the phase and its corresponding color below.

_____

_____

Wire the **stepper motor** (note that another small four-phase stepper motor can be used in this exercise provided that its current demands can be met by the ULN2003A IC) **to the EDE1200 stepper motor interface chip as shown in Figure E13.1**. The ULN2003A chip contains an array of Darlington transistors. It is used here to meet the current demands of the stepper motor, which requires 500 mA per phase since the EDE1200 can only supply a maximum current of 25 mA. The EDE1200 requires an external oscillator connected to pins 15 and 16 to work. The ED1200 can be operated in two modes: step and run modes. In the wiring diagram shown in Figure E13.1, the chip is set to run in step mode, which needs **an external clock signal** (connected to pin 9). The rotational direction and full/half stepping modes are set by the connections to pins 7 and 8, respectively. In this diagram, the motor is set to rotate clockwise using full stepping mode.

Figure E13.1

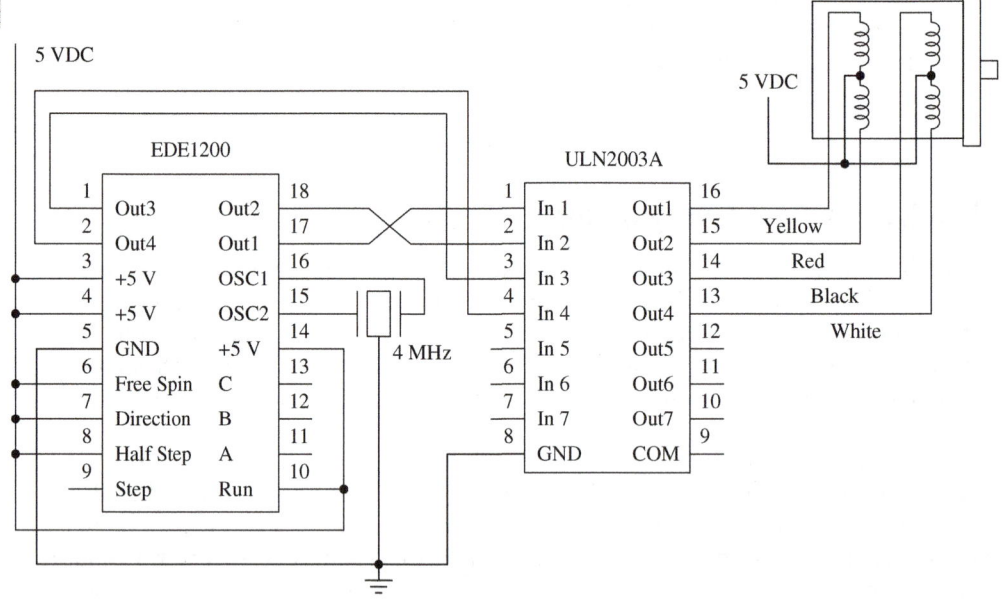

With no power applied to the circuit, rotate the motor shaft by hand. Apply power to the circuit, and rotate the motor shaft again. Comment on what you observed below.

_____

_____

Now attach a small piece of paper (use tape) to the motor shaft so you can observe the rotation of the shaft. Apply power to the circuit. Did the marker move when the power was applied to the circuit? If so, what causes this? Write your answer below.

_____

_____

> **Troubleshooting Tip:** If the circuit is wired correctly, the signals In1, In2, In3, and In4 on the ULN2003A should not be all at the same logic level.

13.2 Use the function generator to **create a square wave** of frequency 1 Hz, V_{max} of 5 V, and a V_{min} of 0 V. **Apply this signal to the step input of the EDE1200 chip** (pin 9). How many steps did it take the motor to do one complete revolution? Write your answer below.

Steps for one revolution _____

Now increase the frequency of the square wave slowly until the motor no longer can rotate. Write down the value of that frequency.

Frequency _____

> **Troubleshooting Tip:** If the motor does not rotate in the direction set by pin 7, then change the order of connecting the motor coils to the ULN2003A.

13.3 Referring to the data sheet of the EDE1200 chip, **change the actuation mode** on the chip **from full stepping to half-stepping mode**. Apply again the 1-Hz step signal to the chip, and determine how many steps it took the motor to do one complete revolution. Write your answer below.

Steps for one revolution _____

Now increase the frequency of the square wave slowly until the motor no longer can rotate. Write down the value of that frequency.

Frequency _____

Is the maximum frequency the same for full and half-stepping modes? Explain below.

_____

_____

_____

13.4* Use the **PIC MCU to move the given motor in full and half-stepping modes** without using the EDE1200 interface chip. In this case, four digital output lines from the MCU are connected directly to the ULN2003A chip, which is in turn connected to the stepper motor. The code inside the MCU should send out timed signals (use a short delay between actuation of the different phase(s)) to drive the four phases of the motor in either full or half-stepping modes. The full or half-stepping mode is set by a digital I/O line that the user sets high or low. Hand in a copy of the code you developed, and demonstrate the operation of your program to your instructor.

> **Hint:** If you are using the low pin-count development board, then you can use the SW1 switch input to select between full and half-stepping.

> **Troubleshooting Tip:** Insure that the PIC development board and the motor supply voltage to the motor share a common ground. If the motor is not rotating, check that the phases are connected in the correct order.

13.5* Redo Exercise 13.4, but do not use the ULN2003A chip. Use **four transistors instead, one for each of the four phases**. Draw a copy of the circuit below. The same code that you developed in Exercise 13.4 should work with no changes needed. Hand in a copy of the code you developed (if you did not do Exercise 13.4) and demonstrate the operation of your program and circuit to your instructor.

QUESTIONS

13.1 Explain how the interface circuit shown in this lab drives the phases of the stepper motor.

13.2 What is the advantage of half-stepping compared to full stepping actuation?

13.3 What determines the current per phase for a stepper motor?

Feedback Control

After you complete this lab, you should be able to:

- Explain the operation of a feedback control system
- Implement a feedback controller in a PIC MCU
- Describe the effects of changing the control gains on system performance

For this lab you need:

- Several 1 kΩ resistors
- Small DC motor with tachometer (Transicoil 1121-110 DC)
- H-bridge (SN754410) or transistor (IRFZ14)
- One diode (1N4004)

▎LAB BACKGROUND/INFORMATION:

A feedback control system is one that tends to maintain a prescribed relationship between the output and the reference input by comparing these and using the difference as a means of control [1]. The components of a feedback control system are the controller, which is the mind of the system; the plant or process to be controlled; and the measuring element or sensor.

The proportional, integral, derivative (PID) controller is one of the most widely used controllers in industry. More than 90% of all industrial controllers are implemented using this popular control law [2]. In continuous time, the controller takes the form

$$Y(t) = K_p e(t) + K_i \int e(t)dt + K_d \frac{d}{dt} e(t)$$

(14.1)

As seen in Equation (14.1), the output of the controller is proportional to the error signal (P-term), the integral of the error signal (I-term), and the derivative of the error signal (D-term) through the gains K_p, K_i, and K_d, respectively.

When using a PC or MCU, a discrete form of the PID controller is implemented. The PID controller is approximated by

$$y(kT) = K_p e(kT) + K_i T \sum_{j=0}^{k-1} e(jT) + K_d((e(kT) - e((k-1)T))/T$$

(14.2)

or alternatively

$$u_i(kT) = u_i((k-1)T) + K_i Te((k-1)T)$$
$$y(kT) = K_p e(kT) + u_i(kT) + K_d((e(kT) - e((k-1)T))/T$$

(14.3)

where T is the sampling interval, u_i is the I-action control output, and k ($k = 0, 1, 2 \ldots$) is an index that represents the number of the instance at which control is done. Notice how the integral for the I-action term is now replaced by a summation and the derivative for the D-action is now replaced by a difference equation.

In this lab, we will utilize a digital version of **a PI controller** to control the speed of a DC motor equipped with a built-in tachometer.

REFERENCES

[1] K. Ogata, *Modern Control Engineering, 4th Edition*, Prentice Hall, 2002.

[2] *The Control Handbook*, Edited by W. Levine, Chapter 10, CRC Press, 1996.

LAB 14 EXERCISES:

Name(s): _____

14.1 The purpose of this exercise is to **perform embedded feedback control of the speed of a DC motor using a PIC MCU**. The given DC motor (see Figure E14.1a) has a built-in tachometer with a sensitivity of 1.9 V/1000 rpm. You are asked to implement a PI or a PID speed controller in this lab. You need to use a timer to perform your control action at a certain timing rate. Timers were covered in Laboratory 10, and you should make use of the timing code given in Exercise 10.1. A block diagram of the operation of this system is shown in Figure E14.1b.

Figure E14.1a

A DC motor with tachometer
(Jouaneh, University of Rhode Island.)

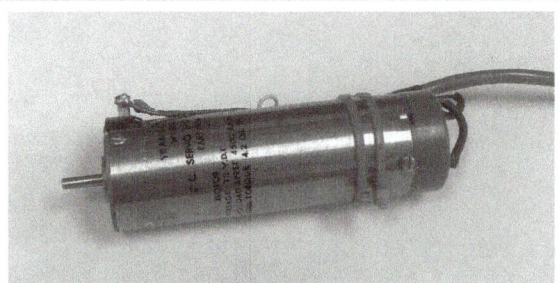

Figure E14.1b

Block diagram of the components of this system

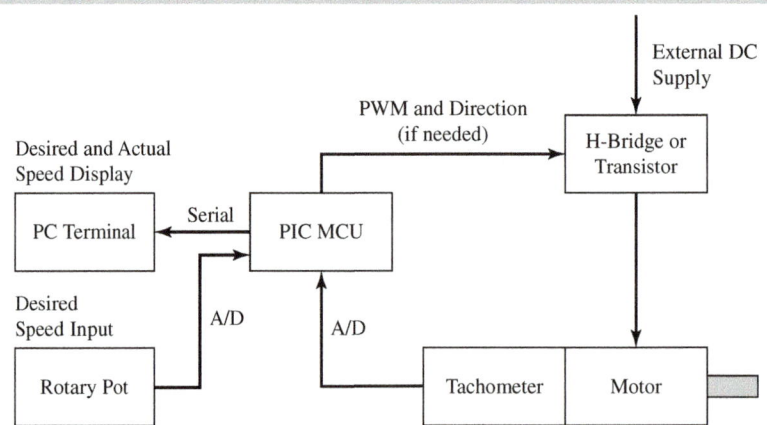

The desired speed of the motor should be read as a 10-bit input on the A/D channel connected to the rotary potentiometer on the Microchip low pin-count board. Because the PIC A/D microcontroller is set to read only positive voltages, the desired speed should be limited to a positive number in the range of 0 to 1023. To vary the set speed, rotate the potentiometer knob on the board. The actual speed of the motor should be read from any of the other available A/D channels on the board.

Hint: You can minimize the ripple in the tachometer output by using a low-pass filter mounted across the tachometer leads, as shown in Figure E14.1a.

Hint: Come up with a way to indicate to the user (through using one or more of the LEDs) that the actual speed is close to the desired speed.

Because the PIC MCU does not have a D/A, you need to use the PWM feature on the PIC microcontroller to send the control output. In this exercise, we will utilize an H-bridge (or a transistor) to interface the PIC MCU to the 12-V power supply to the motor. The control output is varied by changing the duty cycle of the PWM signal. H-bridges were covered in Laboratory 5, while PWM output was covered in Laboratory 8.

To aid in debugging your code and to provide means for displaying the motor speed, utilize the *printf()* function to transmit characters to a terminal program. Use the *PuTTY* program or any another terminal program that is available on the PC. The terminal program communications setting should be set to 38400 baud rate, no parity, 8 data bits, 1 stop bit, and no flow control. Note that sending data often through the serial line will affect the operation of your control program.

> **Hint:** Consider changing the sampling time, and see its effect on the response of the system.

For this exercise, provide a printout of the code you developed. Also, demonstrate the operation of your program to your instructor.

> **Troubleshooting Tip:** You need to place a flyback diode across the motor leads to prevent glitches from the PWM actuation of the motor from affecting the program's operation.

> **Troubleshooting Tip:** If you run your motor at high speeds, then you need to place a voltage-dividing circuit on the tachometer output so that the tachometer output falls within the allowable A/D range for the PIC MCU.

14.2* Develop a VBE program to control the speed of a DC motor tachometer using a digital PI-speed controller. Use the A/D converter on the PC data-acquisition card to read the tachometer signal and the D/A converter to send the controller output to the motor through a servo amplifier or an H-bridge. Use the Performance Counter to implement a *Tsamp* for the system of 1 ms or less. The program should implement a user interface similar to that shown in Figure E14.2.

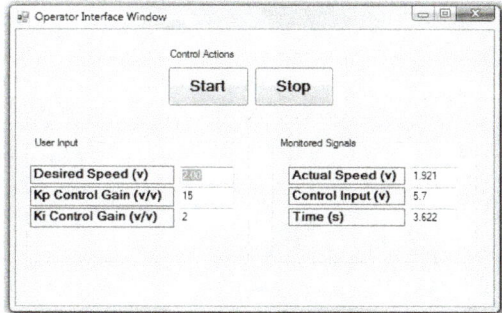

Figure E14.2

The program should use the task state/software structure for implementing the digital controller. For this exercise, provide a printout of the code you developed. Also, demonstrate the operation of your program to your instructor. (*Note*: This exercise assumes the availability of a data-acquisition card with a software library for accessing the A/D and the D/A.)

14.3* Revise the MCU code that you developed in Excercise 14.1 so that the feedback controller is implemented using timer overflow interrupts of Timer0 instead of polling Timer1 to time the control signals. The frequency of the interrupts is controlled by setting the timing parameters for Timer0, since the interrupt occurs when the 8-bit Timer0 overflows. Select the timing parameters (clock frequency and prescaler) to get close to 500 Hz interrupt frequency (or close to 2 ms sampling interval). Test the operation of your program. For this exercise, provide a printout of the code you developed. Also, demonstrate the operation of your program to your instructor.

QUESTIONS

14.1 How did the choice of the sampling time affect the response of the system?

14.2 What happens if one uses a very large control gain?

14.3 Can the response of the system controlled in this lab become oscillatory?

14.4 Did the motor achieve a zero steady-state speed error using the developed controller?

Suggested Projects

This section gives an outline for several project ideas that are suitable for extended or final projects. All of the suggested projects make use of material covered in the previous laboratory exercises. For all of these projects, the student should submit the following.

a. A short report explaining the approach for solving the project.

b. A detailed circuit diagram for interfacing the components in the project.

c. State-transition diagrams for code (if applicable) with all transitions between the states shown and the states in the diagram corresponding exactly to those used in the code.

d. A printout of all of the developed code and screen shots of the code in operation.

e. Electronic copy of the hex file (for MCU projects) and/or contents of bin/Debug folders (for VBE projects) e-mailed to the instructor.

A table of the main components needed for these projects is provided at the end of this section.

SP1. MCU Sensor Monitoring System with PC Display

This project involves using the PIC MCU to perform timed reading of the output of two or more sensors using the different interface devices available on the microcontroller and then to transmit this data through a serial line to the PC that acts as monitor. A block diagram of the operation of this system is shown in Figure SP1.1. The LM35C sensor (see Laboratory 12) provides an analog output voltage, while the DS1631 (see Laboratory 9) provides temperature data in a digital form using the I²C interface.

Figure SP1.1

Block diagram of the sensor monitoring system

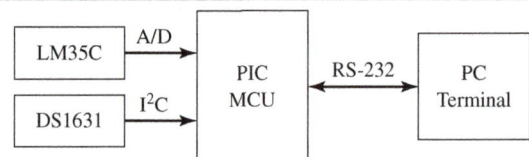

The MCU should read the output from the two sensors at a user specified interval (1 to 5 seconds), convert the output data from each sensor to temperature in engineering units, and transmit the data to a PC that acts as a monitor. To provide flexibility, the MCU will transmit data in one or two forms, depending on the command sent from the PC. The PC commands are the following.

A: This command should cause the MCU to send the average value of the last five readings (in engineering units) obtained from each sensor.

I: This command should cause the MCU to send the individual readings (in engineering units) as they are available with no averaging done.

> **Note:** That two temperature sensors can be replaced by any other sensors. In addition, the terminal interface can be replaced by a VBE GUI which transmits and receives data from the MCU.

SP2. PC CONTROL OF A STEPPER-DRIVEN LINEAR POSITIONING SYSTEM

This project controls the motion of a linear positioning system. The motion system is a stepper-motor driven lead-screw stage operating in open-loop fashion. The stage has two end-of travel limit switches. An example of such a system is shown in Figure SP2.1.

Limit
Switch

Figure SP2.1

An example of a linear positioning system
(Jouaneh, University of Rhode Island.)

We will use the PC for control of this system. A block diagram for the components of this system is shown below. Since we are using a PC, we need to be able to send the step and direction commands to the stage. We assume that the user has access to a data acquisition system in which the step and direction signals can be sent through the parallel digital I/O port to the stepper-motor driver. Similarly, the program also should be able to read the output of the two limit switches using the parallel digital I/O port.

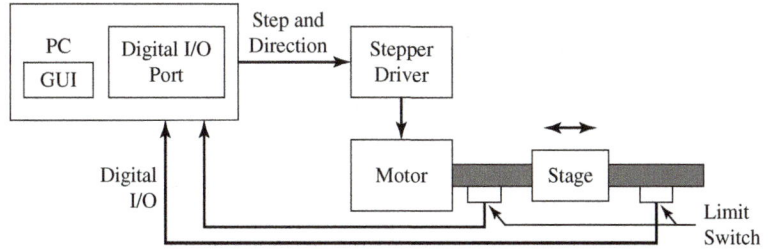

Figure SP2.2

Block diagram of a stepper-driven system

For this project, utilize the same GUI interface and implement the same code structure that was developed in Exercise 11.4, but modify the code to have the ability to control a real stage. As described in Exercise 11.4, your modifications should be limited to the *MotionTask*. The *MotionTask* should send the step and direction output to the stepper-motor driver based on variables that are set in the other task, the *SequenceTask*. To provide flexibility, the software should be able to run the stage in either real or simulated mode. In the real mode, the step and direction signals are sent to the driver, while in simulated mode, the step and direction signals are used to increment a variable that indicates the stage position. The rate of incrementing/decrementing the stage variable position is the same as that for the step rate. Thus, if the user has increased the stage speed in the real system, then the simulated system speed should change accordingly. In a similar fashion, the end-of-travel limit switches are read by a function that returns the state of the real switches if operating in real mode or interprets the simulated stage position variable if operating in simulated mode. For example, the stage is considered to be at the limit switch if the position variable reaches a user-specified limit value (such +/100).

SP3. MCU Control of a DC Linear Actuator System

This project considers the control of a DC linear actuator that consists of a DC motor coupled to an ACME screw. The actuator operates in a closed-loop fashion and uses a linear potentiometer to provide a feedback signal. A photo of the system is shown in Figure SP3.1.

Figure SP3.1

Photo of DC Linear Actuator
(Midwest Motion Products, Watertown, MN.)

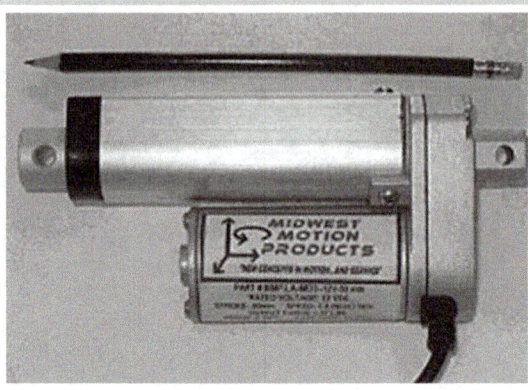

We would like to use the PIC MCU to control the motion of this system. A block diagram of the components of this system is shown in Figure SP3.2. The MCU should implement a simple P or PI feedback controller to move the actuator to three predefined positions: *In*, *Middle*, and *Out*. The *In* and *Out* positions should be close to the travel limit of the actuator but not at the limits. When the actuator reaches the indicated position (or within a close proximity to it), an LED should be lighted to indicate the arrival at the position. To provide motion in both directions, an H-bridge should be used to power up the motor that drives the system. The actuator position (as indicated by the potentiometer) should be read by one of the available A/D channels on the MCU. The output from the feedback controller will set the duty cycle and direction inputs to the H-bridge driver. The desired position is selected from mapping the A/D output of the rotary pot on the MCU development board to these positions, as indicated in the table below.

In	Middle	Out
0–340	341–681	682–1023

To provide feedback on the behavior of this system, the user should use the serial port to display status information about system performance. The feedback controller can be implemented as a timed task (see Laboratory 14). Adjust the controller gains to give a smooth motion.

Figure SP3.2

Block diagram of the DC linear actuator system

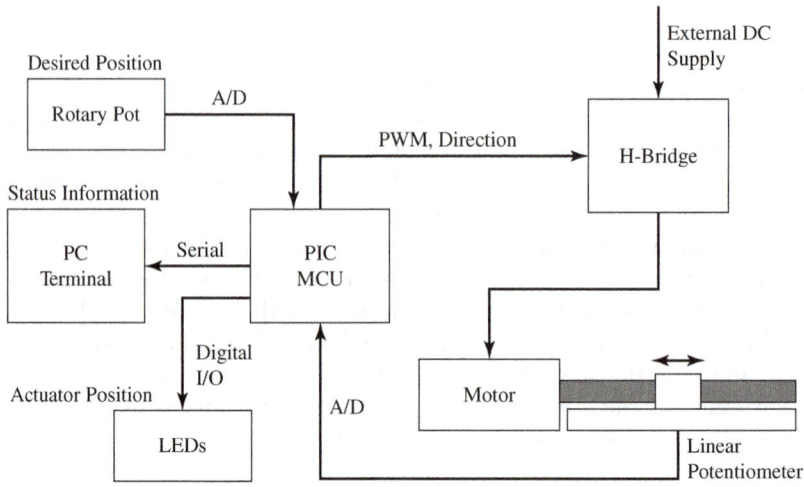

SP4. MCU Oven Simulator with PC Display

This project uses a PIC MCU to simulate the operation of an oven similar to that described in Exercises 11.6 and 11.7. The 'simulated oven' is a small copper plate $(2" \times 1.5" \times 0.5")$ that is heated by a flexible heater that is attached to the bottom of the plate as shown in Figure SP4.1. The plate and heater are mounted on an insulated base. The 'oven temperature' is read from an analog output IC temperature sensor (see Laboratory 12) that is inserted into a small drilled hole in the copper plate. The heater power is modulated by a PWM signal (see Laboratory 8) that is sent to a power transistor (see Laboratory 4). Due to the limited power output of the heater (10 W) and because the suggested temperature sensor has a temperature measurement range of -40 to $110°C$, the simulated oven will have desired oven temperatures that are less than real oven temperatures of 300 to 550°F. Similarly, the 'HOT' indicator of the oven will be set at temperature lower than that used in Exercise 11.6. The user interface should be similar to that described in Exercise 11.7. The simulated oven should operate using the rules listed in Exercise 11.6, but the heating and cooling rates are now determined by dynamics of the plate/heater system.

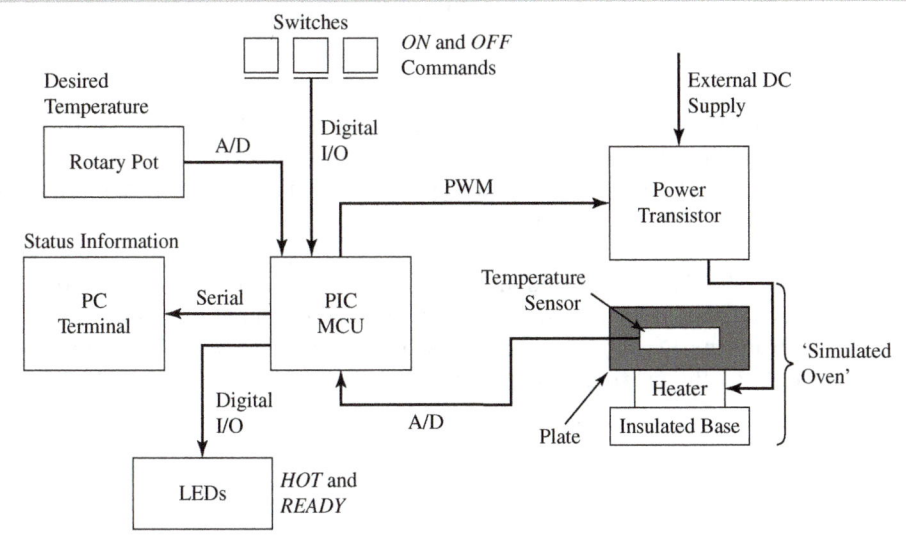

Figure SP4.1

Block diagram of 'simulated oven' system

Note: This project can alternatively be implemented using a PC (No MCU used) provided that the PC has access to a data acquisition card to enable the PC to interface with the physical system.

SP5. Motor Speed Control in a PIC MCU with a PC GUI

In this project, you will extend the motor speed control project that was considered in Laboratory 14 in the following ways.

1. Use VBE to develop a GUI interface that transmits the desired speed and controller gains to the PIC MCU instead of using a potentiometer to set the desired speed and controller gains that are programmed in code. The VBE GUI will also display the actual speed of the motor.

2. Add functionality in the VBE GUI to have the ability to save the data transmitted from the MCU and to write that data to a data file.

3. Add functionality to the code to allow the system to perform open-loop step speed response tests. In this mode, the MCU sends a user-specified voltage signal to the motor (using the PWM) interface and records the speed of the motor over a user-specified time interval. The open-loop step response results allow the user to identify the dynamics of the system, which aids in determining the control gains to be used.

A block diagram of the operation of this system is shown in Figure SP5.1.

Figure SP5.1

Block diagram of motor speed control system

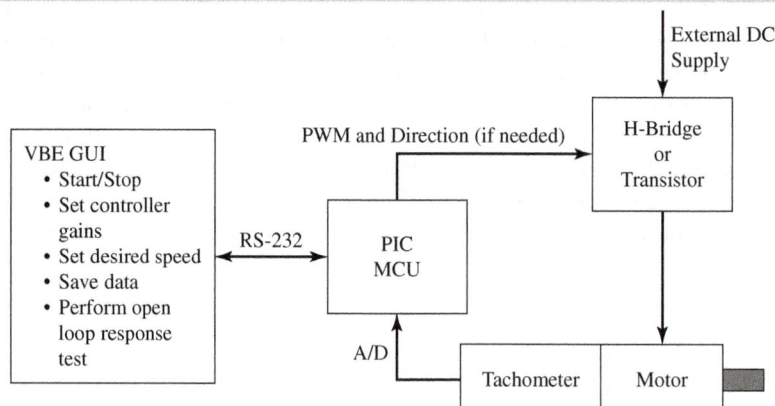

One important aspect of this project is the structure of the control software in both the PC and the MCU and the software issues in transferring data between the two systems. The code in each computing platform should be structured using the task/state control software structure (see Laboratory 11).

Since the PC and the MCU are running independently, a handshaking mechanism also should be employed in the transfer of data between the two programs to ensure that the data is transmitted correctly. In a handshaking mechanism, no new data is sent from the PC to the MCU unless the PC receives an acknowledgement from the MCU on the previous data transfer.

System	Component	Part/Model #	Comments
Sensor Monitoring System	Analog temperature sensor	LM35C	Any temperature sensor with analog output can be used.
	I²C digital temperature Sensor	DS1631	Any temperature sensor with I²C or SPI output can be used.
Stepper Stage	Stepper motor	Oriental Motor PK245-01AA	Other stepper motors that are compatible to this can be used.
	Stepper-motor driver	Applied Motion Products PDO 2035 step motor driver	
	Stage	Velmex XN10-0120-E50-71 slide	The specified stage has 12 in. of travel, a lead of 0.05 in./rev, and has two limit switches. Note that this stage is not the one shown in Figures SP2.1.
DC Linear Actuator	DC linear actuator with built-in linear potentiometer	Midwest Motion Products MMP LA3-12-20-A-200-P	The specified linear actuator uses a 12 VDC supply, has a 20:1 gear ratio, and a 200 mm stroke length.
	H-bridge	SN754410	The current rating of this H-bridge is only adequate to drive the linear actuator with no load applied to the actuator.
Oven Simulator	Heater	McMaster-Carr #7945T52 DC Volt flexible silicone-rubber heat strip adhesive backed, 1" X 2", 10 W	This heater has a 10 W output power value. A heater with a different power output will decrease/increase the time constant of the system.
	Temperature sensor	LM35C sensor	
	Power transistor	IRFZ14 transistor	Any transistor with a power rating of 10 W or higher and a current limit of 1 A can be used.
Motor Speed Control	Motor/w tachometer	Transicoil 1121-110 DC servo motor tachometer from Servo Systems, Inc.	Any PM DC-motor w/tachometer can be used provided that the input and output voltage levels are compatible.
	H-bridge	SN754410	

Table P.1

List of main components for suggested projects

Cable Connectors

Banana Plug	 (Glen Pitt-Pladdy / Shutterstock.)
BNC Connector	 (Albert Lozano / Shutterstock.)
Alligator Clip	 (Bruce MacQueen / Shutterstock.)
Plunger Clip	 (Jouaneh, University of Rhode Island.)
Safety Spade Terminal	 (Jouaneh, University of Rhode Island.)
Molex Connector	 (Courtesy of Molex, Lisle, IL.)

Manufacturer Data Sheets—Front Pages

LM741 Op-amp

 *National Semiconductor*

August 2000

LM741
Operational Amplifier

General Description

The LM741 series are general purpose operational amplifiers which feature improved performance over industry standards like the LM709. They are direct, plug-in replacements for the 709C, LM201, MC1439 and 748 in most applications.

The amplifiers offer many features which make their application nearly foolproof: overload protection on the input and output, no latch-up when the common mode range is exceeded, as well as freedom from oscillations.

The LM741C is identical to the LM741/LM741A except that the LM741C has their performance guaranteed over a 0˚C to +70˚C temperature range, instead of −55˚C to +125˚C.

Features

Connection Diagrams

Metal Can Package

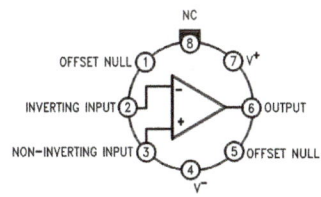

00934102

Note 1: LM741H is available per JM38510/10101

Order Number LM741H, LM741H/883 (Note 1),
LM741AH/883 or LM741CH
See NS Package Number H08C

Ceramic Flatpak

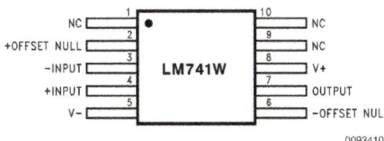

00934106

Order Number LM741W/883
See NS Package Number W10A

Dual-In-Line or S.O. Package

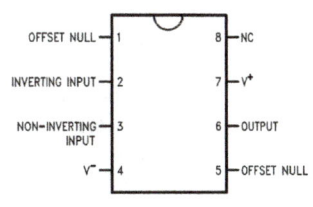

00934103

Order Number LM741J, LM741J/883, LM741CN
See NS Package Number J08A, M08A or N08E

Typical Application

Offset Nulling Circuit

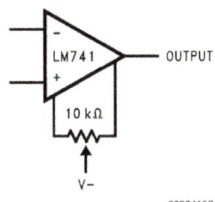

00934107

NMV1215SC DC/DC Converter

NMV 5V, 12V & 15V Series

3kVDC Isolated 1W Single & Dual Output DC/DC Converters

FEATURES

- RoHS compliant
- Efficiency to 79%
- Power density up to 0.85W/cm³
- Wide temperature performance at full 1 Watt load, −40°C to 85°C
- Single or dual output
- UL 94V-0 package material
- No heatsink required
- Footprint from 1.17cm²
- Industry standard pinout
- Power sharing on dual output
- 3kVDC isolation (1 minute)
- 5V, 12V, & 15V input
- 5V, 9V, 12V and 15V output
- Internal SMD construction
- Fully encapsulated with toroidal magnetics
- No external components required
- MTTF up to 4.2 million hours
- No electrolytic or tantalum capacitors

PRODUCT OVERVIEW

The NMV series of industrial temperature range DC/DC converters are the standard building blocks for on-board distributed power systems. They are ideally suited for providing local supplies on control system boards with the added benefit of 3kVDC galvanic isolation to reduce switching noise. Available in SIP and DIP with dual and single output pinout. All of the rated power may be drawn from a single pin provided the total load does not exceed 1 watt.

For full details go to www.murata-ps.com/rohs

SELECTION GUIDE

Order Code	Nominal Input Voltage	Output Voltage	Output Current	Input Current at Rated Load	Load Regulation (Typ)	Load Regulation (Max)	Ripple & Noise (Typ)	Ripple & Noise (Max)	Efficiency	Isolation Capacitance	MTTF[1]	Package Style
	V	V	mA	mA	%		mVp-p		%	pF	kHrs	
NMV0505DAC	5	5	200	294	14.6	15	15	17	68	23	4241	DIP
NMV0509DAC	5	9	111	267	9.3	10	11.3	15	75	30	3376	DIP
NMV0512DAC	5	12	84	260	7.4	8.0	10.5	16	77	26	2555	DIP
NMV0515DAC	5	15	67	256	6.7	7.3	8.7	11	78	27	1838	
NMV0505SAC	5	5	200	294	14.6	15	16	23	68	23	4241	SIP
NMV0509SAC	5	9	111	267	9.3	10	12	15	75	30	3376	SIP
NMV0512SAC	5	12	84	260	7.4	8.0	11	15	77	26	2555	
NMV0515SAC	5	15	67	256	6.7	7.3	11	14	78	27	1838	
NMV1205DAC	12	5	200	121	14.6	15	9.5	14	69	26	2664	DIP
NMV1209DAC	12	9	111	113	9.3	10	7	8.5	74	35	2295	DIP
NMV1212DAC	12	12	84	108	7.4	8.0	8	19	77	43	1883	
NMV1215DAC	12	15	67	108	6.7	7.3	8	17	77	42	1462	
NMV1205SAC	12	5	200	121	14.6	15	11	16	69	26	2664	
NMV1209SAC	12	9	111	113	9.3	10	7.5	14	74	35	2295	
NMV1212SAC	12	12	84	108	7.4	8.0	9	22	77	43	1883	
NMV1215SAC	12	15	67	108	6.7	7.3	8.5	17	77	42	1462	SIP
NMV1505SAC	15	5	200	93	8.3	10	15.5	17	67	21	2747	
NMV1512SAC	15	12	84	85	3.3	4.0	11.2	14	75	45	1365	
NMV1515SAC	15	15	67	84	2.8	4.0	11	13	77	50	941	
NMV0505DC	5	±5	±100	280	9.0	10	11	14	71.5	21	3106	DIP
NMV0509DC	5	±9	±55	263	7.5	8.5	7.5	9	76	24	2258	DIP
NMV0512DC	5	±12	±42	256	6.8	7.5	6.7	9	78	26	1579	
NMV0515DC	5	±15	±33	253	6.8	8.5	6	9	79	27	1065	
NMV0505SC	5	±5	±100	280	9.0	10	11	17	71.5	21	3106	SIP
NMV0509SC	5	±9	±55	263	7.5	8.5	7	9.4	76	24	2258	SIP
NMV0512SC	5	±12	±42	256	6.8	7.5	6.7	8	78	26	1579	
NMV0515SC	5	±15	±33	253	6.8	8.5	6.3	8.2	79	27	1065	
NMV1205DC	12	±5	±100	117	9.0	10	8.6	12	71	27	2148	DIP
NMV1209DC	12	±9	±55	113	7.5	8.5	6.5	9	74	35	1705	DIP
NMV1212DC	12	±12	±42	111	6.8	7.5	6.2	8.5	75	42	1287	
NMV1215DC	12	±15	±33	110	6.8	8.5	5.5	8	76	41	924	
NMV1205SC	12	±5	±100	117	9.0	10	10	13	71	27	2148	
NMV1209SC	12	±9	±55	113	7.5	8.5	8	11	74	35	1705	
NMV1212SC	12	±12	±42	111	6.8	7.5	6	10	75	42	1287	
NMV1215SC	12	±15	±33	110	6.8	8.5	6.5	13	76	41	924	SIP
NMV1505SC	15	±5	±100	91	5.5	10	11	12	69	39	1941	
NMV1512SC	15	±12	±42	87	2.6	3.0	7.5	9	75	68	789	
NMV1515SC	15	±15	±33	84	2.3	3.0	7.5	9	77	84	522	

INPUT CHARACTERISTICS

Parameter	Conditions	Min.	Typ.	Max.	Units
Voltage range	Continuous operation, 5V input types	4.5	5	5.5	V
	Continuous operation, 12V input types	10.8	12	13.2	
	Continuous operation, 15V input types	13.5	15	16.5	
Reflected ripple current			20	40	mA p-p

1. Calculated using MIL-HDBK-217F with nominal input voltage at full load.

All specifications typical at T_A=25°C, nominal input voltage and rated output current unless otherwise specified.

Technical enquiries email: mk@murata-ps.com, tel: +44 (0)1908 615232

2010-09-20   KDC_NMV.G01   Page 1 of 5

(Courtesy of Murata Power Solutions, Inc.)

BPW34 Photodiode

BPW34, BPW34S

Vishay Semiconductors

Silicon PIN Photodiode, RoHS Compliant

94 8583

FEATURES

- Package type: leaded
- Package form: top view
- Dimensions (L x W x H in mm): 5.4 x 4.3 x 3.2
- Radiant sensitive area (in mm^2): 7.5
- High photo sensitivity
- High radiant sensitivity
- Suitable for visible and near infrared radiation
- Fast response times
- Angle of half sensitivity: $\varphi = \pm 65°$
- Lead (Pb)-free component in accordance with RoHS 2002/95/EC and WEEE 2002/96/EC

RoHS
COMPLIANT

DESCRIPTION

BPW34 is a PIN photodiode with high speed and high radiant sensitivity in miniature, flat, top view, clear plastic package. It is sensitive to visible and near infrared radiation.
BPW34S is packed in tubes, specifications like BPW34.

APPLICATIONS

- High speed photo detector

PRODUCT SUMMARY

COMPONENT	I_{ra} (µA)	φ (deg)	$\lambda_{0.1}$ (nm)
BPW34	50	± 65	430 to 1100
BPW34S	50	± 65	430 to 1100

Note

Test condition see table "Basic Characteristics"

ORDERING INFORMATION

ORDERING CODE	PACKAGING	REMARKS	PACKAGE FORM
BPW34	Bulk	MOQ: 3000 pcs, 3000 pcs/bulk	Top view
BPW34S	Tube	MOQ: 1800 pcs, 45 pcs/tube	Top view

Note

MOQ: minimum order quantity

ABSOLUTE MAXIMUM RATINGS

PARAMETER	TEST CONDITION	SYMBOL	VALUE	UNIT
Reverse voltage		V_R	60	V
Power dissipation	$T_{amb} \leq 25\ °C$	P_V	215	mW
Junction temperature		T_j	100	°C
Operating temperature range		T_{amb}	- 40 to + 100	°C
Storage temperature range		T_{stg}	- 40 to + 100	°C
Soldering temperature	$t \leq 3\ s$	T_{sd}	260	°C
Thermal resistance junction/ambient	Connected with Cu wire, 0.14 mm^2	R_{thJA}	350	K/W

Note

$T_{amb} = 25\ °C$, unless otherwise specified

For technical questions, contact: detectortechsupport@vishay.com

Document Number: 81521
Rev. 2.0, 08-Sep-08

(Datasheet courtesy of Vishay Intertechnology, Inc.)

2N3904 Transistor

2N3904   MMBT3904   PZT3904

TO-92

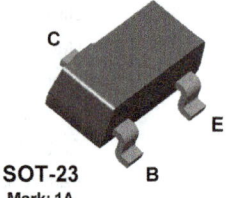

SOT-23
Mark: 1A

SOT-223

NPN General Purpose Amplifier

This device is designed as a general purpose amplifier and switch.
The useful dynamic range extends to 100 mA as a switch and to
100 MHz as an amplifier.

Absolute Maximum Ratings* T_A = 25°C unless otherwise noted

Symbol	Parameter	Value	Units
V_{CEO}	Collector-Emitter Voltage	40	V
V_{CBO}	Collector-Base Voltage	60	V
V_{EBO}	Emitter-Base Voltage	6.0	V
I_C	Collector Current - Continuous	200	mA
T_J, T_{stg}	Operating and Storage Junction Temperature Range	-55 to +150	°C

*These ratings are limiting values above which the serviceability of any semiconductor device may be impaired.

NOTES:
1) These ratings are based on a maximum junction temperature of 150 degrees C.
2) These are steady state limits. The factory should be consulted on applications involving pulsed or low duty cycle operations.

Thermal Characteristics T_A = 25°C unless otherwise noted

Symbol	Characteristic	Max			Units
		2N3904	*MMBT3904	**PZT3904	
P_D	Total Device Dissipation	625	350	1,000	mW
	Derate above 25°C	5.0	2.8	8.0	mW/°C
$R_{\theta JC}$	Thermal Resistance, Junction to Case	83.3			°C/W
$R_{\theta JA}$	Thermal Resistance, Junction to Ambient	200	357	125	°C/W

*Device mounted on FR-4 PCB 1.6" X 1.6" X 0.06."

**Device mounted on FR-4 PCB 36 mm X 18 mm X 1.5 mm; mounting pad for the collector lead min. 6 cm².

(Courtesy of Fairchild Semiconductor, South Portland, ME.)

IRFZ14 Transistor

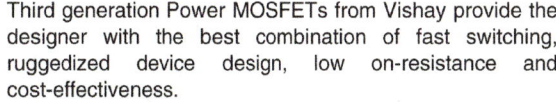

IRFZ14, SiHFZ14

Vishay Siliconix

Power MOSFET

PRODUCT SUMMARY

V_{DS} (V)		60
$R_{DS(on)}$ (Ω)	V_{GS} = 10 V	0.20
Q_g (Max.) (nC)		11
Q_{gs} (nC)		3.1
Q_{gd} (nC)		5.8
Configuration		Single

TO-220

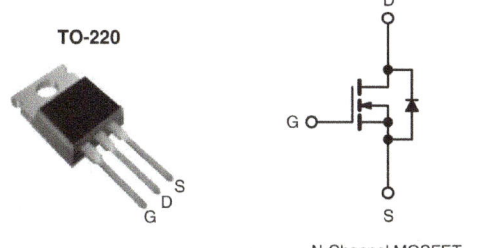

N-Channel MOSFET

FEATURES

- Dynamic dV/dt Rating
- 175 °C Operating Temperature
- Fast Switching
- Ease of Paralleling
- Simple Drive Requirements
- Compliant to RoHS Directive 2002/95/EC

RoHS*
COMPLIANT

Pb
Available

DESCRIPTION

Third generation Power MOSFETs from Vishay provide the designer with the best combination of fast switching, ruggedized device design, low on-resistance and cost-effectiveness.

The TO-220 package is universally preferred for all commercial-industrial applications at power dissipation levels to approximately 50 W. The low thermal resistance and low package cost of the TO-220 contribute to its wide acceptance throughout the industry.

ORDERING INFORMATION

Package	TO-220
Lead (Pb)-free	IRFZ14PbF
	SiHFZ14-E3
SnPb	IRFZ14
	SiHFZ14

ABSOLUTE MAXIMUM RATINGS T_C = 25 °C, unless otherwise noted

PARAMETER			SYMBOL	LIMIT	UNIT
Drain-Source Voltage[f]			V_{DS}	60	V
Gate-Source Voltage[f]			V_{GS}	± 20	
Continuous Drain Current	V_{GS} at 10 V	T_C = 25 °C	I_D	10	A
		T_C = 100 °C		7.2	
Pulsed Drain Current[a]			I_{DM}	40	
Linear Derating Factor				0.29	W/°C
Single Pulse Avalanche Energy[b]			E_{AS}	47	mJ
Maximum Power Dissipation		T_C = 25 °C	P_D	43	W
Peak Diode Recovery dV/dt[c]			dV/dt	4.5	V/ns
Operating Junction and Storage Temperature Range			T_J, T_{stg}	- 55 to + 175	°C
Soldering Recommendations (Peak Temperature)		for 10 s		300[d]	
Mounting Torque		6-32 or M3 screw		10	lbf · in
				1.1	N · m

Notes
a. Repetitive rating; pulse width limited by maximum junction temperature (see fig. 11).
b. V_{DD} = 25 V; starting T_J = 25 °C, L = 1.47 mH, R_g = 25 Ω, I_{AS} = 8 A (see fig. 12).
c. $I_{SD} \leq$ 10 A, dI/dt $\leq$ 90 A/µs, $V_{DD} \leq V_{DS}$, $T_J \leq$ 175 °C.
d. 1.6 mm from case.

Document Number: 91289
S09-1677-Rev. B, 07-Sep-09

www.vishay.com
1

(Datasheet courtesy of Vishay Intertechnology, Inc.)

RPI-352 Photo-interrupter

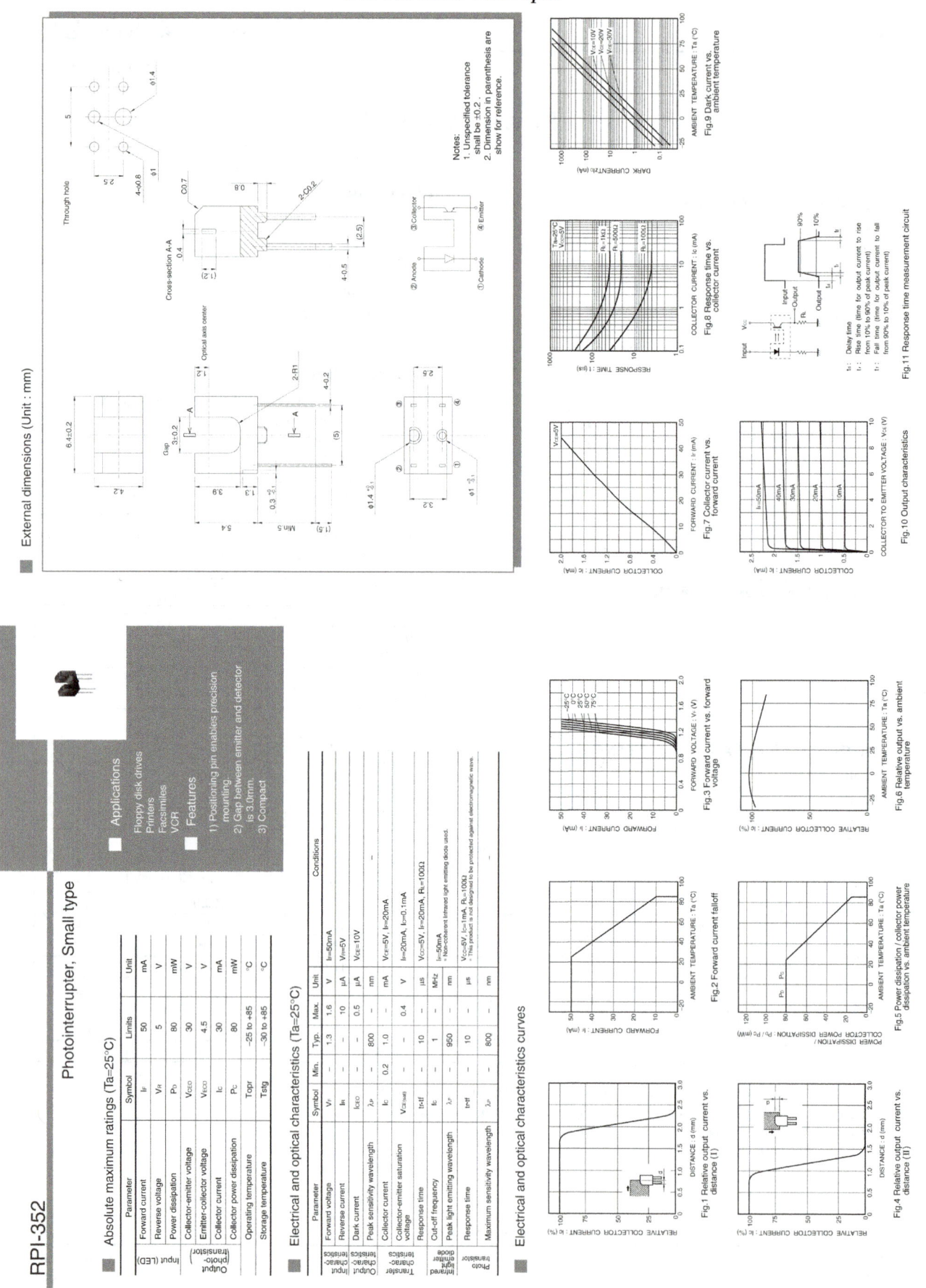

Omron G5V-2 Relay

OMRON

PCB Relay	G5V-2

Miniature Relay for Signal Circuits

- Wide switching power of 10 μA to 2 A.
- High dielectric strength coil-contacts:1,000 VAC; open contacts: 750 VAC.
- Conforms to FCC Part 68 requirements.
- Ag + Au clad bifurcated crossbar contacts and fully sealed for high contact reliability.
- New 150-mW relays with high-sensitivity.

RoHS Compliant   Refer to pages 16 to 17 for details.

℟ ℠ FCC

Ordering Information

Classification	Contact form	Contact type	Contact material	Enclosure ratings	Model
Standard	DPDT	Bifurcated crossbar	Ag + Au-Alloy	Fully sealed	G5V-2
High-sensitivity					G5V-2-H1

Note: When ordering, add the rated coil voltage to the model number.
Example: G5V-2 12 VDC
└── Rated coil voltage

Model Number Legend

G5V - ☐ - ☐   ☐ VDC
     1   2   3

1. **Contact Form**
   2: DPDT

2. **Classification**
   H1: High-sensitivity

3. **Rated Coil Voltage**
   3, 5, 6, 9, 12, 24, 48 VDC

Specifications

■ Coil Ratings

Standard Models

Rated voltage		3 VDC	5 VDC	6 VDC	9 VDC	12 VDC	24 VDC	48 VDC
Rated current		166.7 mA	100 mA	83.3 mA	55.6 mA	41.7 mA	20.8 mA	12 mA
Coil resistance		18 Ω	50 Ω	72 Ω	162 Ω	288 Ω	1,152 Ω	4,000 Ω
Coil inductance	Armature OFF	0.04	0.09	0.16	0.31	0.47	1.98	7.23
(H) (ref. value)	Armature ON	0.05	0.11	0.19	0.49	0.74	2.63	10.00
Must operate voltage		75% max. of rated voltage						
Must release voltage		5% min. of rated voltage						
Max. voltage		120% of rated voltage at 23°C						
Power consumption		Approx. 500 mW						Approx. 580 mW

Note: 1. The rated current and coil resistance are measured at a coil temperature of 23°C with a tolerance of ±10%.
     2. Operating characteristics are measured at a coil temperature of 23°C.
     3. The maximum voltage is the highest voltage that can be imposed on the relay coil.

(Courtesy of Omron Corporation.)

SN754410 H-Bridge

SN754410
QUADRUPLE HALF-H DRIVER

SLRS007B – NOVEMBER 1986 – REVISED NOVEMBER 1995

- 1-A Output-Current Capability Per Driver
- Applications Include Half-H and Full-H Solenoid Drivers and Motor Drivers
- Designed for Positive-Supply Applications
- Wide Supply-Voltage Range of 4.5 V to 36 V
- TTL- and CMOS-Compatible High-Impedance Diode-Clamped Inputs
- Separate Input-Logic Supply
- Thermal Shutdown
- Internal ESD Protection
- Input Hysteresis Improves Noise Immunity
- 3-State Outputs
- Minimized Power Dissipation
- Sink/Source Interlock Circuitry Prevents Simultaneous Conduction
- No Output Glitch During Power Up or Power Down
- Improved Functional Replacement for the SGS L293

NE PACKAGE
(TOP VIEW)

1,2EN	1	16	V_{CC1}
1A	2	15	4A
1Y	3	14	4Y
HEAT SINK AND GROUND	4	13	HEAT SINK AND GROUND
	5	12	
2Y	6	11	3Y
2A	7	10	3A
V_{CC2}	8	9	3,4EN

FUNCTION TABLE
(each driver)

INPUTS†		OUTPUT
A	EN	Y
H	H	H
L	H	L
X	L	Z

H = high-level, L = low-level
X = irrelevant
Z = high-impedance (off)
† In the thermal shutdown mode, the output is in a high-impedance state regardless of the input levels.

description

The SN754410 is a quadruple high-current half-H driver designed to provide bidirectional drive currents up to 1 A at voltages from 4.5 V to 36 V. The device is designed to drive inductive loads such as relays, solenoids, dc and bipolar stepping motors, as well as other high-current/high-voltage loads in positive-supply applications.

All inputs are compatible with TTL-and low-level CMOS logic. Each output (Y) is a complete totem-pole driver with a Darlington transistor sink and a pseudo-Darlington source. Drivers are enabled in pairs with drivers 1 and 2 enabled by 1,2EN and drivers 3 and 4 enabled by 3,4EN. When an enable input is high, the associated drivers are enabled and their outputs become active and in phase with their inputs. When the enable input is low, those drivers are disabled and their outputs are off and in a high-impedance state. With the proper data inputs, each pair of drivers form a full-H (or bridge) reversible drive suitable for solenoid or motor applications.

A separate supply voltage (V_{CC1}) is provided for the logic input circuits to minimize device power dissipation. Supply voltage V_{CC2} is used for the output circuits.

The SN754410 is designed for operation from $-40°C$ to $85°C$.

TEXAS
INSTRUMENTS

(Courtesy of Texas Instruments, Dallas, TX.)

555-Timer

NA555, NE555, SA555, SE555

SLFS022H –SEPTEMBER 1973–REVISED JUNE 2010

PRECISION TIMERS

Check for Samples: NA555, NE555, SA555, SE555

FEATURES

- **Timing From Microseconds to Hours**
- **Astable or Monostable Operation**

- **Adjustable Duty Cycle**
- **TTL-Compatible Output Can Sink or Source up to 200 mA**

NA555...D OR P PACKAGE
NE555...D, P, PS, OR PW PACKAGE
SA555...D OR P PACKAGE
SE555...D, JG, OR P PACKAGE
(TOP VIEW)

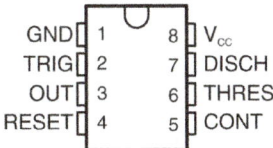

SE555...FK PACKAGE
(TOP VIEW)

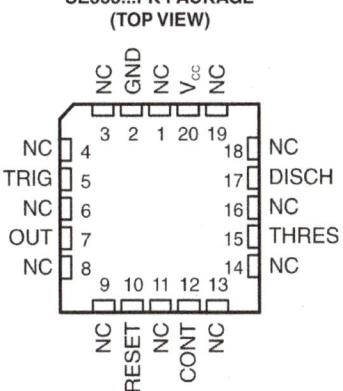

NC – No internal connection

DESCRIPTION/ORDERING INFORMATION

These devices are precision timing circuits capable of producing accurate time delays or oscillation. In the time-delay or monostable mode of operation, the timed interval is controlled by a single external resistor and capacitor network. In the astable mode of operation, the frequency and duty cycle can be controlled independently with two external resistors and a single external capacitor.

The threshold and trigger levels normally are two-thirds and one-third, respectively, of V_{CC}. These levels can be altered by use of the control-voltage terminal. When the trigger input falls below the trigger level, the flip-flop is set, and the output goes high. If the trigger input is above the trigger level and the threshold input is above the threshold level, the flip-flop is reset and the output is low. The reset (RESET) input can override all other inputs and can be used to initiate a new timing cycle. When RESET goes low, the flip-flop is reset, and the output goes low. When the output is low, a low-impedance path is provided between discharge (DISCH) and ground.

The output circuit is capable of sinking or sourcing current up to 200 mA. Operation is specified for supplies of 5 V to 15 V. With a 5-V supply, output levels are compatible with TTL inputs.

Please be aware that an important notice concerning availability, standard warranty, and use in critical applications of Texas Instruments semiconductor products and disclaimers thereto appears at the end of this data sheet.

(Courtesy of Texas Instruments, Dallas, TX.)

AND Gate (SN74LS08)

SN5408, SN54LS08, SN54S08
SN7408, SN74LS08, SN74S08
QUADRUPLE 2-INPUT POSITIVE-AND GATES

SDLS033 – DECEMBER 1983 – REVISED MARCH 1988

- **Package Options Include Plastic ''Small Outline'' Packages, Ceramic Chip Carriers and Flat Packages, and Plastic and Ceramic DIPs**

- **Dependable Texas Instruments Quality and Reliability**

description

These devices contain four independent 2-input AND gates.

The SN5408, SN54LS08, and SN54S08 are characterized for operation over the full military temperature range of −55°C to 125°C. The SN7408, SN74LS08 and SN74S08 are characterized for operation from 0° to 70°C.

FUNCTION TABLE (each gate)

INPUTS		OUTPUT
A	B	Y
H	H	H
L	X	L
X	L	L

logic symbol[†]

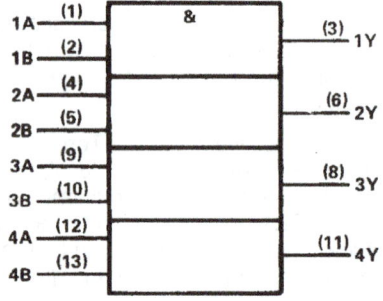

1A — (1)
1B — (2) &
1Y (3)
2A — (4)
2B — (5)
2Y (6)
3A — (9)
3B — (10)
3Y (8)
4A — (12)
4B — (13)
4Y (11)

[†] This symbol is in accordance with ANSI/IEEE Std 91-1984 and IEC Publication 617-12.
Pin numbers shown are for D, J, N, and W packages.

SN5408, SN54LS08, SN54S08 . . . J OR W PACKAGE
SN7408 . . . J OR N PACKAGE
SN74LS08, SN74S08 . . . D, J OR N PACKAGE
(TOP VIEW)

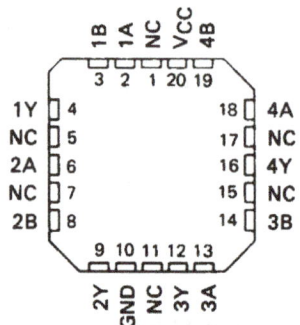

1A	1	14	V$_{CC}$
1B	2	13	4B
1Y	3	12	4A
2A	4	11	4Y
2B	5	10	3B
2Y	6	9	3A
GND	7	8	3Y

SN54LS08, SN54S08 . . . FK PACKAGE
(TOP VIEW)

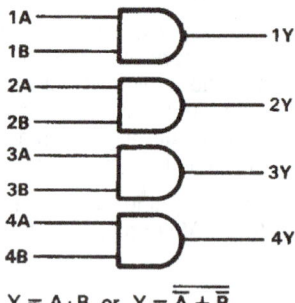

NC—No internal connection

logic diagram (positive logic)

1A
1B
1Y

2A
2B
2Y

3A
3B
3Y

4A
4B
4Y

$$Y = A \cdot B \quad \text{or} \quad Y = \overline{\overline{A} + \overline{B}}$$

TEXAS INSTRUMENTS
POST OFFICE BOX 655303 ● DALLAS, TEXAS 75265

1

(Courtesy of Texas Instruments, Dallas, TX.)

OR Gate (SN74LS32)

SDLS100

**SN5432, SN54LS32, SN54S32,
SN7432, SN74LS32, SN74S32
QUADRUPLE 2-INPUT POSITIVE-OR GATES**

DECEMBER 1983 – REVISED MARCH 1988

- Package Options Include Plastic "Small Outline" Packages, Ceramic Chip Carriers and Flat Packages, and Plastic and Ceramic DIPs
- Dependable Texas Instruments Quality and Reliability

description

These devices contain four independent 2-input OR gates.

The SN5432, SN54LS32 and SN54S32 are characterized for operation over the full military range of −55°C to 125°C. The SN7432, SN74LS32 and SN74S32 are characterized for operation from 0°C to 70°C.

FUNCTION TABLE (each gate)

INPUTS		OUTPUT
A	B	Y
H	X	H
X	H	H
L	L	L

logic symbol [†]

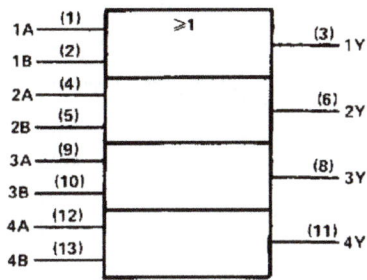

[†] This symbol is in accordance with ANSI/IEEE Std 91-1984 and IEC Publication 617-12.
Pin numbers shown are for D, J, N, or W packages.

SN5432, SN54LS32, SN54S32 . . . J OR W PACKAGE
SN7432 . . . N PACKAGE
SN74LS32, SN74S32 . . . D OR N PACKAGE
(TOP VIEW)

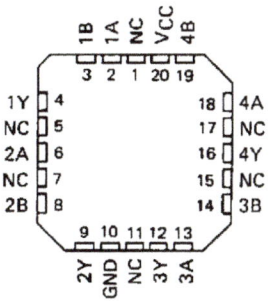

SN54LS32, SN54S32 . . . FK PACKAGE
(TOP VIEW)

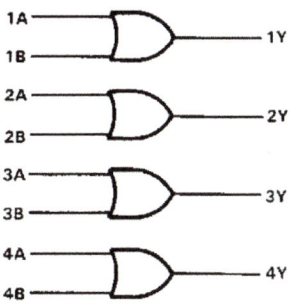

NC - No internal connection

logic diagram

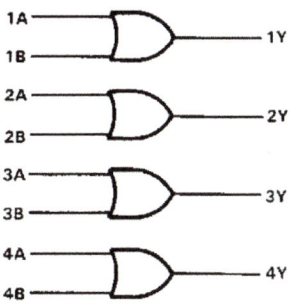

positive logic

$$Y = A + B \text{ or } Y = \overline{\overline{A} \cdot \overline{B}}$$

TEXAS INSTRUMENTS

POST OFFICE BOX 655012 • DALLAS, TEXAS 75265

(Courtesy of Texas Instruments, Dallas, TX.)

Inverter Gate (SN74LS04)

SN5404, SN54LS04, SN54S04, SN7404, SN74LS04, SN74S04
HEX INVERTERS

SDLS029C – DECEMBER 1983 – REVISED JANUARY 2004

● **Dependable Texas Instruments Quality and Reliability**

description/ordering information

These devices contain six independent inverters.

SN5404 . . . J PACKAGE
SN54LS04, SN54S04 . . . J OR W PACKAGE
SN7404, SN74S04 . . . D, N, OR NS PACKAGE
SN74LS04 . . . D, DB, N, OR NS PACKAGE
(TOP VIEW)

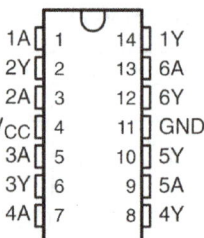

1A	1	14 V_{CC}
1Y	2	13 6A
2A	3	12 6Y
2Y	4	11 5A
3A	5	10 5Y
3Y	6	9 4A
GND	7	8 4Y

SN5404 . . . W PACKAGE
(TOP VIEW)

1A	1	14 1Y
2Y	2	13 6A
2A	3	12 6Y
V_{CC}	4	11 GND
3A	5	10 5Y
3Y	6	9 5A
4A	7	8 4Y

SN54LS04, SN54S04 . . . FK PACKAGE
(TOP VIEW)

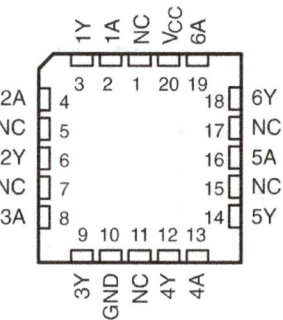

NC – No internal connection

Please be aware that an important notice concerning availability, standard warranty, and use in critical applications of Texas Instruments semiconductor products and disclaimers thereto appears at the end of this data sheet.

TEXAS INSTRUMENTS

POST OFFICE BOX 655303 ● DALLAS, TEXAS 75265

1

(Courtesy of Texas Instruments, Dallas, TX.)

SR Flip-Flop (SN74LS279A)

SN54279, SN54LS279A, SN74279, SN74LS279A
QUADRUPLE S̄-R̄ LATCHES

SDLS093 – DECEMBER 1983 – REVISED MARCH 1988

- Package Options Include Plastic "Small Outline" Packages, Ceramic Chip Carriers and Flat Packages, and Plastic and Ceramic DIPs

- Dependable Texas Instruments Quality and Reliability

description

The '279 offers 4 basic S̄-R̄ flip-flop latches in one 16-pin, 300-mil package. Under conventional operation, the S̄-R̄ inputs are normally held high. When the S̄ input is pulsed low, the Q output will be set high. When R̄ is pulsed low, the Q output will be reset low. Normally, the S̄-R̄ inputs should not be taken low simultaneously. The Q output will be unpredictable in this condition.

SN54279, SN54LS279A . . . J OR W PACKAGE
SN74279 . . . N PACKAGE
SN74LS279A . . . D OR N PACKAGE
(TOP VIEW)

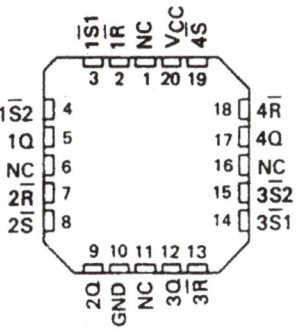

SN54LS279A . . . FK PACKAGE
(TOP VIEW)

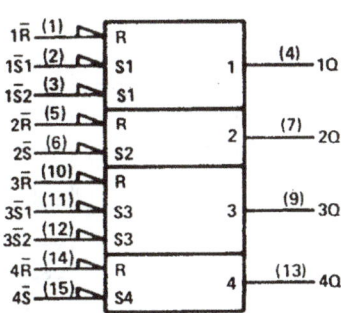

NC - No internal connection

FUNCTION TABLE
(each latch)

INPUTS		OUTPUT
S̄†	R̄	Q
H	H	Q_0
L	H	H
H	L	L
L	L	H‡

H = high level   L = low level

†For latches with double S inputs:
Q_0 = the level of Q before the indicated input conditions were established.

‡ This configuration is nonstable: that is, it may not persist when the S̄ and R̄ inputs return to their inactive (high) level.
H = both S̄ inputs high
L = one or both S̄ inputs low

logic symbol§

§This symbol is in accordance with ANSI/IEEE Std. 91-1984 and IEC Publication 617-12.
Pin numbers shown are for D, J, N, and W packages.

logic diagram (positive logic)

(latches 1 and 3)   **(latches 2 and 4)**

Copyright © 1988, Texas Instruments Incorporated

POST OFFICE BOX 655303 ● DALLAS, TEXAS 75265

(Courtesy of Texas Instruments, Dallas, TX.)

BCD-to-7 Segments Decoder

CD54HC4511, CD74HC4511, CD74HCT4511
BCD-TO-7 SEGMENT LATCH/DECODER/DRIVERS

SCHS279D – DECEMBER 1998 – REVISED OCTOBER 2003

- 2-V to 6-V V_{CC} Operation ('HC4511)
- 4.5-V to 5.5-V V_{CC} Operation (CD74HCT4511)
- High-Output Sourcing Capability
  - 7.5 mA at 4.5 V (CD74HCT4511)
  - 10 mA at 6 V ('HC4511)
- Input Latches for BCD Code Storage
- Lamp Test and Blanking Capability
- Balanced Propagation Delays and Transition Times
- Significant Power Reduction Compared to LSTTL Logic ICs
- 'HC4511
  - High Noise Immunity, N_{IL} or N_{IH} = 30% of V_{CC} at V_{CC} = 5 V
- CD74HCT4511
  - Direct LSTTL Input Logic Compatibility, V_{IL} = 0.8 V Maximum, V_{IH} = 2 V Minimum
  - CMOS Input Compatibility, $I_I \leq 1\ \mu A$ at V_{OL}, V_{OH}

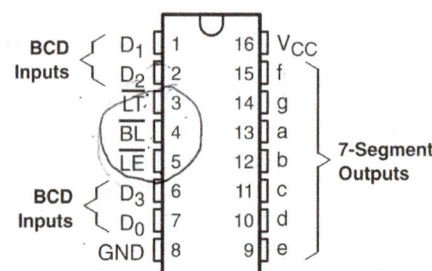

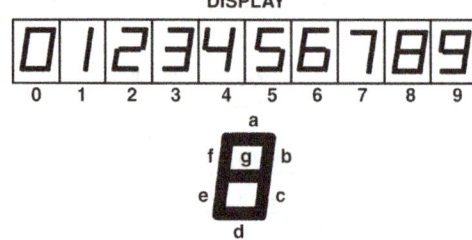

description/ordering information

The CD54HC4511, CD74HC4511, and CD74HCT4511 are BCD-to-7 segment latch/decoder/drivers with four address inputs (D_0–D_3), an active-low blanking ($\overline{BL}$) input, lamp-test ($\overline{LT}$) input, and a latch-enable ($\overline{LE}$) input that, when high, enables the latches to store the BCD inputs. When $\overline{LE}$ is low, the latches are disabled, making the outputs transparent to the BCD inputs.

These devices have standard-size output transistors, but are capable of sourcing (at standard V_{OH} levels) up to 7.5 mA at 4.5 V. The HC types can supply up to 10 mA at 6 V.

ORDERING INFORMATION

T_A	PACKAGE†		ORDERABLE PART NUMBER	TOP-SIDE MARKING
−55°C to 125°C	PDIP – E	Tube of 25	CD74HC4511E	CD74HC4511E
			CD74HCT4511E	CD74HCT4511E
	SOIC – M	Tube of 40	CD74HC4511M	HC4511M
		Reel of 2500	CD74HC4511M96	
		Reel of 250	CD74HC4511MT	
	TSSOP – PW	Reel of 2000	CD74HC4511PWR	HJ4511
		Reel of 250	CD74HC4511PWT	
	CDIP – F	Tube of 25	CD54HC4511F3A	CD54HC4511F3A

† Package drawings, standard packing quantities, thermal data, symbolization, and PCB design guidelines are available at www.ti.com/sc/package.

Please be aware that an important notice concerning availability, standard warranty, and use in critical applications of Texas Instruments semiconductor products and disclaimers thereto appears at the end of this data sheet.

(Courtesy of Texas Instruments, Dallas, TX.)

7-Segments Display

HDSP-311x/313x

10.16 mm (0.4 inch) Single Digit General Purpose
Seven Segment Display

Data Sheet

Description

This 10.16 mm (0.4 inch) LED single digit seven
segment display uses industry standard size package
and pinout. The device is available in either common
anode or common cathode. The choice of colors
includes High Efficiency Red (HER), Green, AlGaAs
Red, and Yellow. The gray face displays are suitable for
indoor use.

Applications

* **Suitable for indoor use**

* **Not recommended for industrial application, i.e.,
  operating temperature requirements exceeding +85˚C
  or below –25˚C[1]**

* **Extreme temperature cycling not recommended**

Note:
1. For additional details, please contact your local Avago sales office or
   an authorized distributor.

Features

* **Industry standard size**

* **Industry standard pinout**
  10.16 mm (0.4 inch) character height
  DIP lead on 2.54 mm

* **Choice of colors**
  High Efficiency Red (HER), Green, AlGaAs Red, and
  Yellow

* **Excellent appearance**
  Evenly lighted segments gray package gives optimum
    contrast
  ± 50 ft. viewing angle

* **Design flexibility**
  Common anode right hand decimal point or common
  cathode right hand decimal point

* **Categorized for luminous intensity**
  Green and yellow categorized for color

Devices

HER	Green	AlGaAs Red	Yellow	Description	Package Drawing
HDSP-311E	HDSP-311G	HDSP-311A	HDSP-311Y	Common Anode Right Hand Decimal	A
HDSP-313E	HDSP-313G	HDSP-313A	HDSP-313Y	Common Cathode Right Hand Decimal	B

(Courtesy of AVAGO Technologies.)

Microchip 25LC256 SPI EEPROM

MICROCHIP

25AA256/25LC256

256K SPI Bus Serial EEPROM

Device Selection Table

Part Number	Vcc Range	Page Size	Temp. Ranges	Packages
25LC256	2.5-5.5V	64 Byte	I, E	P, SN, SM, ST, MF
25AA256	1.8-5.5V	64 Byte	I	P, SN, SM, ST, MF

Features:

- Max. Clock 10 MHz
- Low-Power CMOS Technology:
  - Max. Write Current: 5 mA at 5.5V, 10 MHz
  - Read Current: 6 mA at 5.5V, 10 MHz
  - Standby Current: 1 μA at 5.5V
- 32,768 x 8-bit Organization
- 64-Byte Page
- Self-Timed Erase and Write Cycles (5 ms max.)
- Block Write Protection:
  - Protect none, 1/4, 1/2 or all of array
- Built-In Write Protection:
  - Power-on/off data protection circuitry
  - Write enable latch
  - Write-protect pin
- Sequential Read
- High Reliability:
  - Endurance: 1,000,000 erase/write cycles
  - Data retention: > 200 years
  - ESD protection: > 4000V
- Temperature Ranges Supported:
  - Industrial (I):   -40°C to +85°C
  - Automotive (E):   -40°C to +125°C
- Pb-Free and RoHS Compliant

Pin Function Table

Name	Function
$\overline{CS}$	Chip Select Input
SO	Serial Data Output
$\overline{WP}$	Write-Protect
Vss	Ground
SI	Serial Data Input
SCK	Serial Clock Input
$\overline{HOLD}$	Hold Input
Vcc	Supply Voltage

Description:

The Microchip Technology Inc. 25AA256/25LC256 (25XX256*) are 256 Kbit Serial Electrically Erasable PROMs. The memory is accessed via a simple Serial Peripheral Interface (SPI) compatible serial bus. The bus signals required are a clock input (SCK) plus separate data in (SI) and data out (SO) lines. Access to the device is controlled through a Chip Select ($\overline{CS}$) input.

Communication to the device can be paused via the hold pin ($\overline{HOLD}$). While the device is paused, transitions on its inputs will be ignored, with the exception of Chip Select, allowing the host to service higher priority interrupts.

The 25XX256 is available in standard packages including 8-lead PDIP and SOIC, and advanced packaging including 8-lead DFN and 8-lead TSSOP.

Package Types (not to scale)

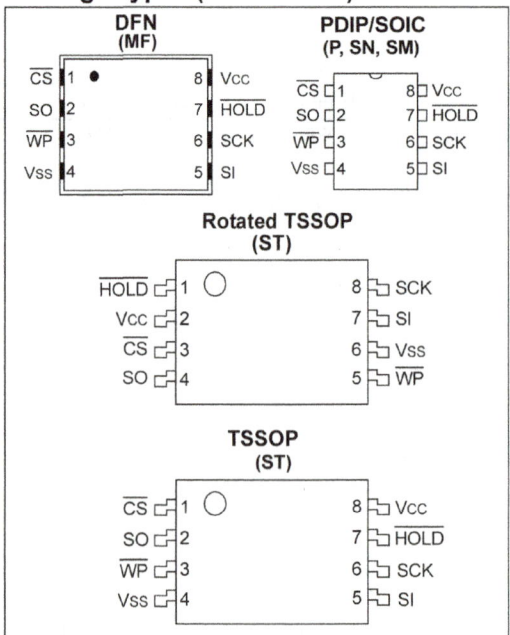

* 25XX256 is used in this document as a generic part number for the 25AA256, 25LC256 devices.

(Reprinted with the permission of Microchip Technology Incorporated.)

DS1631 Temperature Sensor

DS1631/DS1631A/DS1731
High-Precision Digital
Thermometer and Thermostat

www.maxim-ic.com

FEATURES
- DS1631 and DS1631A Provide ±0.5°C Accuracy over a 0°C to +70°C Range
- DS1731 Provides ±1°C Accuracy over a -10°C to +85°C Range
- DS1631A Automatically Begins Taking Temperature Measurements at Power-Up
- Operating Temperature Range: -55°C to +125°C (-67°F to +257°F)
- Temperature Measurements Require No External Components
- Output Resolution is User-Selectable to 9, 10, 11, or 12 Bits
- Wide Power-Supply Range (+2.7V to +5.5V)
- Converts Temperature-to-Digital Word in 750ms (max)
- Multidrop Capability Simplifies Distributed Temperature-Sensing Applications
- Thermostatic Settings are User-Definable and Nonvolatile (NV)
- Data is Read/Written Through 2-Wire Serial Interface (SDA and SCL Pins)
- All Three Devices are Available in 8-Pin µSOP Packages and the DS1631 is Also Available in a 150mil SO package—see Table 1 for Ordering Information

PIN CONFIGURATIONS

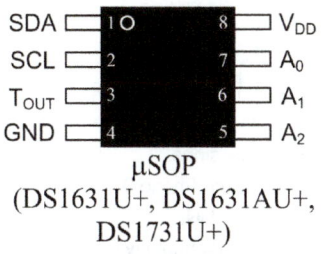

µSOP
(DS1631U+, DS1631AU+, DS1731U+)

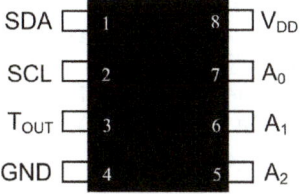

SO (150mil and 208mil)
(DS1631Z+, DS1631S+)

See Table 2 for Pin Descriptions

APPLICATIONS
- Network Routers and Switches
- Cellular Base Stations
- Portable Products
- Any Space-Constrained Thermally Sensitive Product

DESCRIPTION
The DS1631, DS1631A, and DS1731 digital thermometers provide 9, 10, 11, or 12-bit temperature readings over a -55°C to +125°C range. The DS1631 and DS1631A thermometer accuracy is ±0.5°C from 0°C to +70°C with $3.0V \le V_{DD} \le 5.5V$, and the DS1731 accuracy is ±1°C from -10°C to +85°C with $3.0V \le V_{DD} \le 5.5V$. The thermostat on all three devices provides custom hysteresis with user-defined trip points (T_H and T_L). The T_H and T_L registers and thermometer configuration settings are stored in NV EEPROM so they can be programmed prior to installation. In addition, the DS1631A automatically begins taking temperature measurements at power-up, which allows it to function as a stand-alone thermostat. Communication with the DS1631/DS1631A/DS1731 is achieved through a 2-wire serial interface, and three address pins allow up to eight devices to be multidropped on the same 2-wire bus.

Pin descriptions for the DS1631/DS1631A/DS1731 are provided in Table 2 and user-accessible registers are summarized in Table 3. A functional diagram is shown in Figure 1.

102307

MCP42010-I/P Digital Potentiometer

MICROCHIP

MCP41XXX/42XXX

Single/Dual Digital Potentiometer with SPI™ Interface

Features

- 256 taps for each potentiometer
- Potentiometer values for 10 kΩ, 50 kΩ and 100 kΩ
- Single and dual versions
- SPI™ serial interface (mode 0,0 and 1,1)
- ±1 LSB max INL & DNL
- Low power CMOS technology
- 1 µA maximum supply current in static operation
- Multiple devices can be daisy-chained together (MCP42XXX only)
- Shutdown feature open circuits of all resistors for maximum power savings
- Hardware shutdown pin available on MCP42XXX only
- Single supply operation (2.7V - 5.5V)
- Industrial temperature range: -40°C to +85°C
- Extended temperature range: -40°C to +125°C

Block Diagram

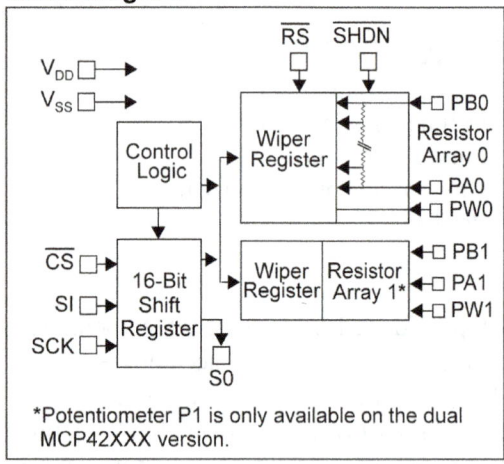

*Potentiometer P1 is only available on the dual MCP42XXX version.

Description

The MCP41XXX and MCP42XXX devices are 256-position, digital potentiometers available in 10 kΩ, 50 kΩ and 100 kΩ resistance versions. The MCP41XXX is a single-channel device and is offered in an 8-pin PDIP or SOIC package. The MCP42XXX contains two independent channels in a 14-pin PDIP, SOIC or TSSOP package. The wiper position of the MCP41XXX/42XXX varies linearly and is controlled via an industry-standard SPI interface. The devices consume <1 µA during static operation. A software shutdown feature is provided that disconnects the "A" terminal from the resistor stack and simultaneously connects the wiper to the "B" terminal. In addition, the dual MCP42XXX has a $\overline{SHDN}$ pin that performs the same function in hardware. During shutdown mode, the contents of the wiper register can be changed and the potentiometer returns from shutdown to the new value. The wiper is reset to the mid-scale position (80h) upon power-up. The $\overline{RS}$ (reset) pin implements a hardware reset and also returns the wiper to mid-scale. The MCP42XXX SPI interface includes both the SI and SO pins, allowing daisy-chaining of multiple devices. Channel-to-channel resistance matching on the MCP42XXX varies by less than 1%. These devices operate from a single 2.7 - 5.5V supply and are specified over the extended and industrial temperature ranges.

Package Types

PDIP/SOIC

MCP41XXX

$\overline{CS}$	1	8 V_{DD}
SCK	2	7 PB0
SI	3	6 PW0
V_{SS}	4	5 PA0

PDIP/SOIC/TSSOP

MCP42XXX

$\overline{CS}$	1	14 V_{DD}
SCK	2	13 SO
SI	3	12 $\overline{SHDN}$
V_{SS}	4	11 $\overline{RS}$
PB1	5	10 PB0
PW1	6	9 PW0
PA1	7	8 PA0

(Reprinted with the permission of Microchip Technology Incorporated.)

MAX233 RS-232 Driver/Receiver

19-4323; Rev 16; 7/10

∧∧XI∧∧

+5V-Powered, Multichannel RS-232 Drivers/Receivers

MAX220–MAX249

General Description

The MAX220–MAX249 family of line drivers/receivers is intended for all EIA/TIA-232E and V.28/V.24 communications interfaces, particularly applications where ±12V is not available.

These parts are especially useful in battery-powered systems, since their low-power shutdown mode reduces power dissipation to less than 5μW. The MAX225, MAX233, MAX235, and MAX245/MAX246/MAX247 use no external components and are recommended for applications where printed circuit board space is critical.

Applications

Portable Computers

Low-Power Modems

Interface Translation

Battery-Powered RS-232 Systems

Multidrop RS-232 Networks

AutoShutdown and UCSP are trademarks of Maxim Integrated Products, Inc.

Next-Generation Device Features

♦ **For Low-Voltage, Integrated ESD Applications**
MAX3222E/MAX3232E/MAX3237E/MAX3241E/MAX3246E: +3.0V to +5.5V, Low-Power, Up to 1Mbps, True RS-232 Transceivers Using Four 0.1μF External Capacitors (MAX3246E Available in a UCSP™ Package)

♦ **For Low-Cost Applications**
MAX221E: ±15kV ESD-Protected, +5V, 1μA, Single RS-232 Transceiver with AutoShutdown™

Ordering Information

PART	TEMP RANGE	PIN-PACKAGE
MAX220CPE+	0°C to +70°C	16 Plastic DIP
MAX220CSE+	0°C to +70°C	16 Narrow SO
MAX220CWE+	0°C to +70°C	16 Wide SO
MAX220C/D	0°C to +70°C	Dice*
MAX220EPE+	-40°C to +85°C	16 Plastic DIP
MAX220ESE+	-40°C to +85°C	16 Narrow SO
MAX220EWE+	-40°C to +85°C	16 Wide SO
MAX220EJE	-40°C to +85°C	16 CERDIP
MAX220MJE	-55°C to +125°C	16 CERDIP

+Denotes a lead(Pb)-free/RoHS-compliant package.
*Contact factory for dice specifications.

Ordering Information continued at end of data sheet.

Selection Table

Part Number	Power Supply (V)	No. of RS-232 Drivers/Rx	No. of Ext. Caps	Nominal Cap. Value (μF)	SHDN & Three-State	Rx Active in SHDN	Data Rate (kbps)	Features
MAX220	+5	2/2	4	0.047/0.33	No	—	120	Ultra-low-power, industry-standard pinout
MAX222	+5	2/2	4	0.1	Yes	—	200	Low-power shutdown
MAX223 (MAX213)	+5	4/5	4	1.0 (0.1)	Yes	✔	120	MAX241 and receivers active in shutdown
MAX225	+5	5/5	0	—	Yes	✔	120	Available in SO
MAX230 (MAX200)	+5	5/0	4	1.0 (0.1)	Yes	—	120	5 drivers with shutdown
MAX231 (MAX201)	+5 and +7.5 to +13.2	2/2	2	1.0 (0.1)	No	—	120	Standard +5/+12V or battery supplies; same functions as MAX232
MAX232 (MAX202)	+5	2/2	4	1.0 (0.1)	No	—	120 (64)	Industry standard
MAX232A	+5	2/2	4	0.1	No	—	200	Higher slew rate, small caps
MAX233 (MAX203)	+5	2/2	0	—	No	—	120	No external caps
MAX233A	+5	2/2	0	—	No	—	200	No external caps, high slew rate
MAX234 (MAX204)	+5	4/0	4	1.0 (0.1)	No	—	120	Replaces 1488
MAX235 (MAX205)	+5	5/5	0	—	Yes	—	120	No external caps
MAX236 (MAX206)	+5	4/3	4	1.0 (0.1)	Yes	—	120	Shutdown, three state
MAX237 (MAX207)	+5	5/3	4	1.0 (0.1)	No	—	120	Complements IBM PC serial port
MAX238 (MAX208)	+5	4/4	4	1.0 (0.1)	No	—	120	Replaces 1488 and 1489
MAX239 (MAX209)	+5 and +7.5 to +13.2	3/5	2	1.0 (0.1)	No	—	120	Standard +5/+12V or battery supplies; single-package solution for IBM PC serial port
MAX240	+5	5/5	4	1.0	Yes	—	120	DIP or flatpack package
MAX241 (MAX211)	+5	4/5	4	1.0 (0.1)	Yes	—	120	Complete IBM PC serial port
MAX242	+5	2/2	4	0.1	Yes	✔	200	Separate shutdown and enable
MAX243	+5	2/2	4	0.1	No	—	200	Open-line detection simplifies cabling
MAX244	+5	8/10	4	1.0	No	—	120	High slew rate
MAX245	+5	8/10	0	—	Yes	✔	120	High slew rate, int. caps, two shutdown modes
MAX246	+5	8/10	0	—	Yes	✔	120	High slew rate, int. caps, three shutdown modes
MAX247	+5	8/9	0	—	Yes	✔	120	High slew rate, int. caps, nine operating modes
MAX248	+5	8/8	4	1.0	Yes	✔	120	High slew rate, selective half-chip enables
MAX249	+5	6/10	4	1.0	Yes	✔	120	Available in quad flatpack package

∧∧XI∧∧

Maxim Integrated Products 1

For pricing, delivery, and ordering information, please contact Maxim Direct at 1-888-629-4642, or visit Maxim's website at www.maxim-ic.com.

LM35C Temperature Sensor

N *ational* **S** *emiconductor*

November 2000

(right margin, vertical) **LM35 Precision Centigrade Temperature Sensors**

LM35
Precision Centigrade Temperature Sensors

General Description

The LM35 series are precision integrated-circuit temperature sensors, whose output voltage is linearly proportional to the Celsius (Centigrade) temperature. The LM35 thus has an advantage over linear temperature sensors calibrated in ° Kelvin, as the user is not required to subtract a large constant voltage from its output to obtain convenient Centigrade scaling. The LM35 does not require any external calibration or trimming to provide typical accuracies of ±¼°C at room temperature and ±¾°C over a full −55 to +150°C temperature range. Low cost is assured by trimming and calibration at the wafer level. The LM35's low output impedance, linear output, and precise inherent calibration make interfacing to readout or control circuitry especially easy. It can be used with single power supplies, or with plus and minus supplies. As it draws only 60 µA from its supply, it has very low self-heating, less than 0.1°C in still air. The LM35 is rated to operate over a −55° to +150°C temperature range, while the LM35C is rated for a −40° to +110°C range (−10° with improved accuracy). The LM35 series is available pack-

aged in hermetic TO-46 transistor packages, while the LM35C, LM35CA, and LM35D are also available in the plastic TO-92 transistor package. The LM35D is also available in an 8-lead surface mount small outline package and a plastic TO-220 package.

Features

■ Calibrated directly in ° Celsius (Centigrade)
■ Linear + 10.0 mV/°C scale factor
■ 0.5°C accuracy guaranteeable (at +25°C)
■ Rated for full −55° to +150°C range
■ Suitable for remote applications
■ Low cost due to wafer-level trimming
■ Operates from 4 to 30 volts
■ Less than 60 µA current drain
■ Low self-heating, 0.08°C in still air
■ Nonlinearity only ±¼°C typical
■ Low impedance output, 0.1 Ω for 1 mA load

Typical Applications

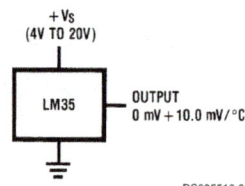

DS005516-3

FIGURE 1. Basic Centigrade Temperature Sensor
(+2°C to +150°C)

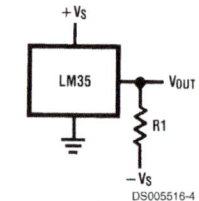

DS005516-4

Choose $R_1 = -V_S/50 \ \mu A$
$V_{OUT} = +1,500$ mV at +150°C
   $= +250$ mV at +25°C
   $= -550$ mV at −55°C

FIGURE 2. Full-Range Centigrade Temperature Sensor

MP101301 Magnetic Proximity Sensor

MAGNETIC SENSOR

MP1013 Series

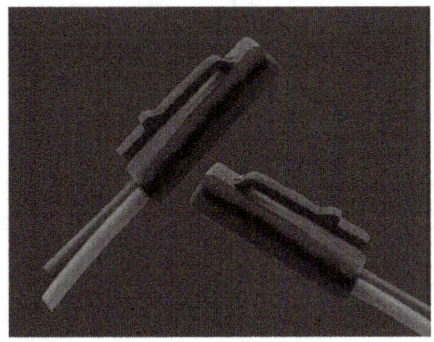

Hall-effect proximity sensor with convenient snap-fit mounting.

Features
- Solid State Reliability
- Excellent output stability over operating temperature range
- Regulated power supply not required
- Meets IEC529 IP67 for dust and water protection
- Open Collector (NPN) output can be used with bipolar switch or cmos logic circuits with suitable pull up resistor

MP101301 and MP101302 – unipolar switch
- Output switches low (off) when the magnetic field at the sensor exceeds the operate point threshold.
- Output switches high (on) when the magnetic field is reduced to below the release point threshold

MP101303 – bipolar latch
- Output latches high(on) in the presence of a south pole.
- Output unlatches (low or off) in the presence of a north pole

Applications:
- Speed sensing
- Door interlock sensing
- Water flow sensing

Specifications

Part Number	Operating Voltage Range (VDC)	Supply Current (mA max.)	Output	Output Saturation Voltage (mV max.)	Output Current (mA max.)	Operating Temp Range (°C)	Storage Temp Range (°C)	Operate Point Gauss (max.)	Release Point Gauss (min.)	Leads	Reverse Battery Protection
MP101301	4.75 – 24	9	3-wire sink	400	25	-40 to 85	-40 to 105	300	60	24 AWG x 150mm	-24VDC
MP101302	4.75 – 24	9	3-wire sink	400	25	-40 to 125	-40 to 125	300	60	24 AWG x 150mm	-24VDC
MP101303	3.5 – 24	4	3-wire sink	500	25	-40 to 85	-40 to 105	45	-45 (latch)	24 AWG x 150mm	None

Notes: These sensors require the use of an external pull-up resistor, the value of which is dependent on the supply voltage.

Recommended pull-up resistor values:					
Volts DC	5	9	12	15	24
Ohms	1K	1.8K	2.4K	3K	3K

Dimensions inches (mm)

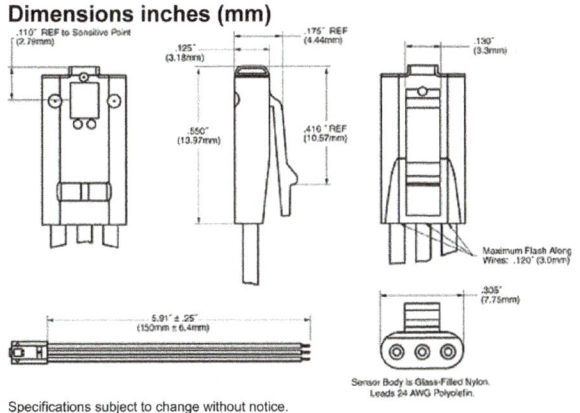

Sensor Body is Glass-Filled Nylon.
Leads 24 AWG Polyolefin.

Specifications subject to change without notice.

Open Collector Sinking Block Diagram

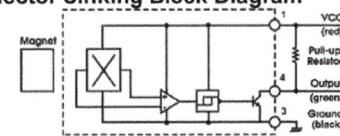

Sensor Pocket

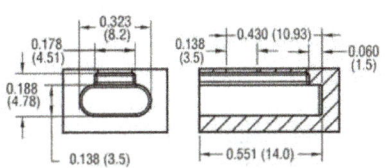

Revised 0111

EDE1200 Unipolar Stepper IC

Jameco Part Number 141532

EDE1200 Unipolar Stepper Motor IC

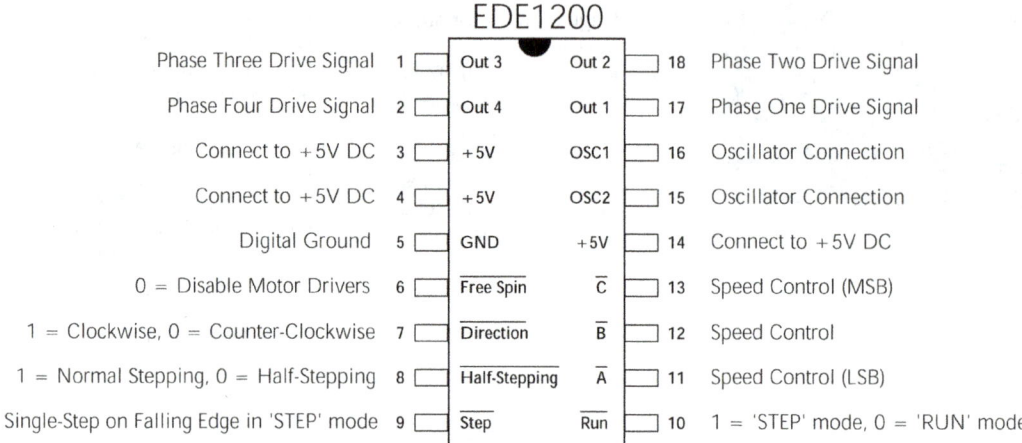

The EDE1200 Unipolar Stepper Motor IC is a 5 volt, 18 pin package designed to interface a logic-level input byte to a stepper motor. The EDE1200 is capable of self-clocking in the free-standing 'RUN' mode, as well as external clocking in the 'STEP' mode. In addition, half-stepping and directional control are also available. The TTL-level outputs sequence the stepper drive circuits that consist of standard power transistors or a transistor array IC. The EDE1200 features the ability to change the stepping rate while the motor is stepping and to take an unlimited number of steps in continuous 'RUN' mode. Inputs are TTL/ CMOS compatible.

RUN mode

In the 'RUN' mode, activated by a low on pin 10, the EDE1200 will cause the motor to rotate according to the following parameters:

Direction (pin 7): 1 = clockwise,   0 = counter-clockwise
(If a clockwise command causes counterclockwise rotation of motor, reverse the sequence of the motor's four phase wires.)

Half-Stepping (pin 8): 1 = normal stepping,   0 = half stepping (doubles step resolution)

Speed Control [C,B,A] (pins 13,12,11): these three active-low bits select one of eight rotational speeds. Refer to Tables One & Two below for speed range details.

(Courtesy of E-Lab Digital Engineering, Inc., Independence, MO. For EDE1200 see PDN1200 at www.paladinsemi.com.)

ULN2003A Darlington Array

TEXAS INSTRUMENTS

**ULN2002A, ULN2003A, ULN2003AI, ULN2004A
ULQ2003A, ULQ2004A**

www.ti.com

SLRS027J–DECEMBER 1976–REVISED JUNE 2010

HIGH-VOLTAGE, HIGH-CURRENT DARLINGTON TRANSISTOR ARRAYS

Check for Samples: ULN2002A, ULN2003A, ULN2003AI, ULN2004A, ULQ2003A, ULQ2004A

FEATURES

- **500-mA-Rated Collector Current (Single Output)**
- **High-Voltage Outputs: 50 V**
- **Output Clamp Diodes**
- **Inputs Compatible With Various Types of Logic**
- **Relay-Driver Applications**

ULN2002A . . . N PACKAGE
ULN2003A . . . D, N, NS, OR PW PACKAGE
ULN2004A . . . D, N, OR NS PACKAGE
ULQ2003A, ULQ2004A . . . D OR N PACKAGE
(TOP VIEW)

1B	1	16	1C
2B	2	15	2C
3B	3	14	3C
4B	4	13	4C
5B	5	12	5C
6B	6	11	6C
7B	7	10	7C
E	8	9	COM

DESCRIPTION

The ULN2002A, ULN2003A, ULN2003AI, ULN2004A, ULQ2003A, and ULQ2004A are high-voltage high-current Darlington transistor arrays. Each consists of seven npn Darlington pairs that feature high-voltage outputs with common-cathode clamp diodes for switching inductive loads. The collector-current rating of a single Darlington pair is 500 mA. The Darlington pairs can be paralleled for higher current capability. Applications include relay drivers, hammer drivers, lamp drivers, display drivers (LED and gas discharge), line drivers, and logic buffers. For 100-V (otherwise interchangeable) versions of the ULN2003A and ULN2004A, see the SN75468 and SN75469, respectively.

The ULN2001A is a general-purpose array and can be used with TTL and CMOS technologies. The ULN2002A is designed specifically for use with 14-V to 25-V PMOS devices. Each input of this device has a Zener diode and resistor in series to control the input current to a safe limit. The ULN2003A and ULQ2003A have a 2.7-kΩ series base resistor for each Darlington pair for operation directly with TTL or 5-V CMOS devices. The ULN2004A and ULQ2004A have a 10.5-kΩ series base resistor to allow operation directly from CMOS devices that use supply voltages of 6 V to 15 V. The required input current of the ULN/ULQ2004A is below that of the ULN/ULQ2003A, and the required voltage is less than that required by the ULN2002A.

Please be aware that an important notice concerning availability, standard warranty, and use in critical applications of Texas Instruments semiconductor products and disclaimers thereto appears at the end of this data sheet.

(Courtesy of Texas Instruments, Dallas, TX.)